Holy People, Holy IrReverence

A Church in Need of Reform and Renewal

By

Rev. David L. Harpe

authorHOUSE™

1663 LIBERTY DRIVE, SUITE 200
BLOOMINGTON, INDIANA 47403
(800) 839-8640
WWW.AUTHORHOUSE.COM

First published by AuthorHouse 5/25/07

ISBN: 1-4184-9779-7 (sc)

Printed in the United States of America
Bloomington, Indiana

This book is printed on acid-free paper.

For all who follow the way of Jesus Christ, ours is not an easy path to travel. While the journey is dotted with experiences of great joy and exuberance, it is also marked by moments of pain, bewilderment and sadness. But these are all part of one and the same journey, which Christians make not only as individuals, but also as a **community of believers.** Thus as a people on their way, our experiences of joy and exuberance are to be shared and celebrated by all. In the same way, it is together that we are most able to effectively acknowledge, confront and ultimately move beyond the occasional moments of pain, bewilderment and sadness.

Something of this shared journey is given expression in the pages of this book, especially as pertains to those of us who seek to follow the Christ from within the tradition of Roman Catholicism. Within these pages, the reader will encounter a people who have known great joy, a people who are presently enduring much sadness, frustration and even anger, especially as reflected in some of their own words and the words of this writer. But further, the reader will encounter a people of tremendous hope.

Pray for us, during this present moment of darkness, that we may find healing and newness of life, that we may faithfully return to the path of Christ. Pray that these words of St. Francis of Assisi will, again, become our own:

We have been called to heal wounds, to unite what has fallen apart and to bring home those who have lost their way.

May God richly bless the journey of each and every one of us within the church and throughout the world.

Fr. David L. Harpe

Table of Contents

THE LIGHT SHINES IN THE DARKNESS, AND THE
DARKNESS CANNOT OVERCOME IT.

-John 1 : 5

This book is dedicated to the people of the now-closed Saint
Joseph Catholic Church, Muskegon, Michigan. Theirs was a rich
history and an incredible story of faith. In the four short years that
I was privileged to journey with this people, the Evangelist's words
were confirmed time and time again: "and the darkness cannot
overcome it."

Thank you, people of St. Joseph, for being a gift of immeasurable
value in my personal journey of faith, and for being that vehicle
through which The Light, Jesus Christ, continues to shine.

With love and prayers......always,

Fr. Dave

ABOUT THIS BOOK

Long before my ordination to the priesthood in 1994, I had come to believe that the Catholic Church was in critical need of reform and renewal. This is not some new or private revelation, but rather an affirmation of the church's teaching that the people of God are always in the process of ongoing conversion, formation, and renewal in faith as we journey together toward the Kingdom of Heaven. My experience as a priest, along with the shared experiences of a growing number of Catholics, has served only to underscore this belief that we are a church in need of reform and renewal. Adding to this the continued unfolding of the clergy sex abuse scandal, one can only conclude that we the church, and our leaders in particular, must once again embrace our own church teaching, perhaps now more than ever in recent history.

In July of 1997 I arrived at St. Joseph Parish. There I would spend four years in my first assignment as a pastor. Little did any of us know, for we couldn't know, that I would also be the parish's last pastor. This aside, much happened during those four years and beyond. And it is the entirety of the experience, which, as stated in the dedication, has become that gift of immeasurable value. It was shortly after my arrival at St. Joe's that I began to entertain thoughts of writing a book of sorts about our church's need for change, reform and renewal. Initially, it was my desire to write solely from a pastor's perspective. However, I soon realized that this book must also be written out of a parish's lived experience. After all, this is about **our** journey, theirs and mine together. As such, what follows is the telling of that journey which, in many ways, is shared by Catholics beyond the walls of St. Joseph, even beyond any diocesan boundaries. Within our journey there were highs and lows, unexpected turns and challenges, laughter and tears. But through it all, God has remained faithful.

It is my privilege to share this story in writing. Chapters 1 and 2 are written from the experience of the last four years in the life of a parish. And while the majority of Chapter 3 is primarily a sharing of this priest's personal journey, it reflects the experience of Catholics who continue to call or write in a concert of voices: "This is not just your story, or the story of the people of St. Joseph. This is **my** story as well. And it needs to be told."

With great joy, then, I share our story. It is written in simple language and, at the surface level, covers a period of roughly six years. As the people of St. Joe's have been a gift to me, so I hope that our story will be something of a gift to the larger church. For it is apparent that we are not alone.

ABOUT THE TITLE

Having spoken of the church's need for reform and renewal, a few words may be called for with regard to "holy people" and "holy irreverence." So as to be very clear, the latter has nothing to do with the scandal of clergy sex abuse, which is not at all "holy," and which can only be described as shameful, sinful, criminal and immoral, to say the least.

Two words, "holy" and "sacred," are often used synonymously. While they enjoy an intimate connection, it is important to note the subtle difference between them. For the purpose of this writing, a theological distinction seems most appropriate.

Perhaps you have heard this expression: "A symbol is a sign, but a sign is not a symbol." What's the difference? A sign stands as an indicator or marker, giving direction and pointing to something beyond itself. A symbol, in theological language, not only points to something beyond itself, but also participates in the very object or reality to which it is signaling. In similar fashion, to refer to something or someone as "holy" is to suggest a sharing or participation in that which is signified, namely, the sacred (i.e., the divine, God).

It is this way of thinking about ourselves as church, which, in part, led to the titling of this book. As a holy people, sharing in the very life of the divine, we are also a people who are on our way. A work in progress, the journey of life and faith is one of continued growth in holiness, of growing more fully into the divine, of which we are already an image (see Genesis 1 : 26-27) and which is already within us. Thus, "holy people" is a wonderfully spiritual and biblical expression of our calling, our identity, and even our ultimate destination.

"Holy Irreverence" came to be included in the title of this book as something of a playful expression, which also finds its roots in Sacred Scripture. Throughout the biblical texts, this pattern emerges: Priests, Prophets, Kings and the like are anointed (chosen) by God and given the task of shepherding God's people. When they fail in their calling and begin to neglect or bring harm to the people entrusted to their care, God anoints yet another as their replacement. As is often the case, God is willing to take risks along the way. So too, those who are anointed by God are called to be, among other things, risk-takers, all for the sake of stretching Gods' people and calling them to deeper growth in the life of faith. Many a priest, prophet and king have come to know, first-hand, the inherent risks and danger which seem to come with the territory when one says "yes" to God. At the very least, that person may find him / herself subjected to scorn and ridicule, or told that they are unholy and irreverent. And at its worst, that individual may have to pay the ultimate price for their faithfulness to God and their willingness to take some risks **with** God.

At St. Joe's we dared to take some risks. Pastor and parishioners were, for the most part, open and willing to challenge and stretch one another in faith. Fortunately, none were called upon to pay that ultimate price. However, some of our neighboring and pre-Vatican II Catholic brothers and sisters were very quick to call us "unholy" and "irreverent," especially when the word went out, for example, that during Liturgy the people of St. Joseph were standing for the Eucharistic prayer, or that families were taking turns making the unleavened bread for weekend liturgies. And so it is out of this, the experience of the last four years in the life of St. Joseph Parish and the continued journey and struggle within myself and many Catholics, that I have playfully and deliberately titled this book HOLY PEOPLE, HOLY IRREVERENCE: A Church in Need of Reform and Renewal.

Chapter One
ST. JOSEPH

A Brief History and Background

Over a period of many years, beginning in the mid-1800's, there came to be nine Catholic churches within the larger community of Muskegon. A number of the earlier parishes emerged in response to the needs of Catholic immigrants who were settling into the area and hoping to build new lives in this country. Within greater Muskegon this would have included German, French, Italian, Hungarian, Irish and Polish Catholics. The smallest of these faith communities was that of St. Joseph. Established in 1883, St. Joseph was home to Muskegon's German Catholics. Under the dedicated leadership of its early pastors, the parish thrived as a house of faith and a gathering space for the German people as they learned a new language and became part of mainstream America. As was often the pattern for many parishes, a Catholic school and a convent would follow, becoming part of the landscape and life of St. Joseph.

In the normal flow of life it is only natural that time and circumstance have a role to play in the human drama. These impact our lived experience, indeed our very lives, whether for better or worse. Nothing is exempt from their influence, including the structures and institutions that rise to become an integral part of that lived experience. Ultimately, all life and all things succumb to time and circumstance. In the span of some forty-plus years, the people of St. Joseph would come to a first-hand encounter with this truth. And for many, myself included, the experience of the last few years would become a most sacred moment in our journey of faith.

The Beginning of the End of a Parish

To better appreciate the events of the final years in the life of St. Joe's, one must first have a general sense of the larger community and its experience. Because I am not a historian, I will take the liberty of writing in a manner by which some artists apply paint to a canvas, that is, with very broad strokes.

The early twentieth century saw Muskegon transformed from "Lumber Town," as it was known, to an industrial city that was home to a paper mill and a significant number of factories and foundries. As such, Muskegon became a place of growing opportunity. The industrial boom attracted many to Lake Michigan's "port city." Among them were African Americans. Arriving primarily from Detroit, Chicago and the deep south, this people came to Muskegon in search of that which was drawing so many others, namely, opportunity for the present and the hope of a brighter future. For some within the African American community this hope, this dream would be slowly and painfully realized. For others, their search continues, even to this day.

In many respects, Muskegon was not unlike any other industrial town or city. Growth in any way, shape or form is often accompanied by growing pain, which presents its own challenges. And along the way, depending on how the community embraces itself and especially its challenges, there are moments of both light and darkness. One significant challenge for Muskegon had to do with numbers.

It was only natural for a large number of Muskegon's new arrivals, especially African Americans, to look for living space within the existing residential neighborhoods that encompassed the greater downtown area. I say "natural" because many of these new arrivals, lacking the personal luxury of an automobile, found it necessary to live in close proximity to their places of employment. From the various neighborhoods, they could then rely on the city's bus system to meet their basic transportation needs.

As Muskegon's industry grew, so too did its population. While the downtown area was becoming more congested, the surrounding neighborhoods were fast approaching capacity. And this was the challenge for Muskegon, that of accommodation. Muskegon, like many other cities, responded to this challenge in the way of low-income housing. Initially, this was one of those moments of light for the larger community and a real boon for many who were simply looking to find a home. At the same time, however, there was something of a grassroots response to Muskegon's growing population. For Muskegon, as it was for many cities nationwide, this particular response would be one of those darker moments along the way. To this day we have termed that response: WHITE FLIGHT.

In the spirit of fairness, many whites moved away from the inner-city neighborhoods in direct response to growing congestion and overcrowding. But in the spirit of truth, there is no denying that many whites also moved away from the same neighborhoods in direct response to the growing number of blacks who were moving in. "White flight" is more than a dark moment. Then and now (and it does continue), the phenomenon of while flight underscores the darker side of humanity. Then and now, it highlights the reality of fear, ignorance and hatred. Indeed, it shows itself to be among the darkest of that which we call SIN.

Decline

What does all the above have to do with St. Joseph? The outward movement of whites and, to some degree, their abandonment of the inner-city neighborhoods resulted in a process of decline that had a direct and negative impact not only to St. Joseph, but to other downtown parishes as well. As white Catholics moved out and suburban life emerged and flourished, suburban parishes would likewise be established. Parishes such as St. Mary, St. Jean Baptiste and St. Joseph would be especially hard-hit by the effects of time and circumstance. With the establishment of two parishes in 1948 and another in 1975, the downtown parishes lost large numbers of parishioners. And with the loss of parishioners came the loss of revenue. As things changed, as people moved away, so too did the downtown parishes change. With their congregations becoming smaller, comprised primarily of aged and loyal parishioners, the effects of time and circumstance seemed to signal, for many, the beginning of the end.

This certainly appeared to be the case for St. Joseph. The parish was both a witness and casualty of time and circumstance. It watched as people moved out of the Nelson Neighborhood, which was home to St. Joe's for nearly 118 years. It watched as the neighborhood became racially mixed. It watched as the neighborhood, over a period of years, wrestled with a rise in crime and drug-related activity. And it watched, during its last years, as the neighborhood began to take ownership of itself in an effort to make a comeback. In and through all of this, however, the parish itself continued to decline. But this period of decline was not without its moments of light, life and blessing.

Historically, as previously stated, St. Joseph was the smallest of Muskegon's Catholic parishes. During its last thirty years or more, however, this distinction came to be noted with such regularity and even a sense of pride that it seemed to be the very heart of the parish's identity: St. Joseph, the smallest of our parishes. That it was. But it was the heart and spirit of the parish that made St. Joseph something big

It was heart and spirit that enabled the parish to survive years of declining membership and financial struggle. It was heart and spirit that sustained the parish as it lived for decades in the shadow of the threat of closure. It was heart and spirit that led the parish to not only welcome change, but to be among the first to advocate for change, which often involves taking risks. And it was heart and spirit which gave to St. Joseph, the smallest of our parishes, the gift of being the most diverse of Muskegon's Catholic faith communities: We were of Black, White, Hispanic and Native American ancestry. There was affluence as well as poverty. But all of this would disappear when the faith community gathered as one in the presence of our God. And it was precisely this experience, in its entirety, which led many to refer to St. Joe's as "the best-kept secret in Muskegon."

Heart, spirit, soul, faith. All of these are essential elements of any faith community. But we cannot forget that critical element of time and circumstance. Despite its heart, spirit and all that sustained the parish along the way, its numbers continued to steadily decline. Time and circumstance were catching up with St. Joseph, the smallest of our parishes.

At The Edge

It was the first week of July, 1997, that I arrived at St. Joseph. A priest for only three years, I was filled with excitement and anticipation as I prepared to receive my first pastorate. But there was also present within me a small measure of fear and anxiety. How would the people of St. Joseph receive me? And with the parish having been served by a string of fine pastors, would I be up to following in the steps of such priests as Fathers Irmen, Hasenberg, Stratz, Hall and Johnson?

The fear and anxiety disappeared rather quickly, almost instantly. For you see, the smallest of our parishes was blest with a very large heart. This parish, this people were gifted with a lively spirit of warmth, welcome and hospitality. I had heard of this trademark spirit long before my arrival. Now, it was mine to experience first-hand. I was welcomed in, immediately feeling a sense of belonging. And as I've said to many persons, even to this day, "I was at home long before my bags were unpacked."

The first order of business as a new and first-time pastor was that of allowing a minimum six-month period of transition, both for myself and for the people of St. Joe's. This was to be first a time for connecting, for getting to know one another. It was also a time for me to simply observe, with an eye toward the presence of some of those elements that are basic, yet critical to the life of every parish. For example:

- Is there life within this particular community when it comes together in prayer?
- What kinds of numbers (of persons) are gathering regularly for Liturgy?
- Content and quality of liturgical celebrations
- Liturgical music

This period of connecting and observing proved to be most beneficial. The smallness of the parish was among its greatest assets, in that it made for a very intimate congregation whose members were clearly well-connected to one another. There was no mistaking the intimacy of St. Joe's, an observation that was consistently echoed by visitors as well as newcomers to the parish. It was indeed very easy to connect, to come to know the people of St. Joseph and to be "at home" with them.

About those basic yet critical elements of a parish, the period of observation yielded much. There was, without question, an abundance of good. At the same time, it was apparent that there were areas in which the parish was lacking. This was not a bad thing, but rather a simple reminder that St. Joseph was much like any other faith community. We were NOT a perfect society. We WERE, however, a pilgrim people, journeying together. As such, there is the built-in reminder that we are always immersed in the ongoing process of conversion, reform and renewal in faith, a process that is complete only when we die into the Kingdom.

So, what specifically did my observations yield? Perhaps most striking was the connectedness and liveliness of the people of St. Joe's. As a person who enjoys the use of imagery and analogy, the people of this parish exemplified St. Paul's analogy of the church as one body with many parts (I Corinthians 12). Week in and week out, this people took time to greet and converse with one another as part of our ritual gathering. Often, there was laughter among the adults, accompanied by the pleasant noise of our children. For many Catholics of time past, and for some still today, such "noise" might be viewed as disrespectful and quite possibly irreverent. But for myself and others, this experience was truly wonderful. The very act of people gathering at St. Joe's provided one with a visible image of Paul's analogy: One body, with many parts. We had size, shape, color and diversity. And most important, we were alive. This parish, unlike many others, had more than just a pulse. To encounter this people was to have an encounter with the living, breathing body of Christ.

Beneath the trademark warmth, hospitality and liveliness of this people, St. Joseph, there lay a deep and solid faith. This, too, was readily recognizable. When the noise and laughter quieted, and the community formally entered into the Liturgy, a collective spirit remained. I wondered, privately, if the parish would be able to articulate this about themselves, but they possessed a great sense of the communal nature of Liturgy. Theirs was very much the action of a people, not a room full of individuals praying.

For the most part, this period of observation showed forth a people who were quite healthy in their faith, but not necessarily without need. Again, as individuals and as a people we are always in need. While the people of St. Joseph gathered with deep faith and a lively spirit, liturgical celebrations appeared somewhat dull. This

I attributed primarily to the lack of good liturgical music, which is perhaps the single most important element in helping to build and shape a people's liturgical experience.

Despite the parish's liveliness of spirit and depth of faith, there were a couple of issues of practical concern that were equally recognizable. One had to do with numbers. The large majority of parishioners were at or beyond retirement age, which often means fixed and/or limited income for themselves and ultimately the parish. The other concern, as just alluded to, was that of parish income. Obviously, these go hand-in-hand, one effecting the other.

In a nutshell, what I observed during my first six months at St. Joseph was a people who were very much alive in their faith. But I also observed that the process of decline was continuing. With every death, new life was not coming into the parish. There were gaping holes in the crowd at both weekend liturgies, and weekly parish income was in the neighborhood of $1,200-$1,400. Privately, I began to wonder why the parish had remained open. Publicly, however, I began to speak to a congregation about time and circumstance catching up with the parish, and that we were now finding ourselves at the edge.

Taking Risks

By the end of the six-month period of observation I found myself speaking very candidly to a congregation about our need to be risk-takers. Given the parish's history and the continuing process of decline in its numbers, the decision to ask a parish to go out on a limb was really a matter of survival. Our options were limited: Either we dare to try something different, or we continue down the current path. The choice before us seemed to be, quite literally, life or death.

The decision to ask a parish to consider risk and change is not to be entered into lightly. Among Catholics one will find some who balk at the slightest hint of change. At the same time, one will find those for whom change cannot come fast enough. These two groups represent something of the extreme. But in between is the larger group, the bulk of the parish, which tends to bring a kind of balance to the whole. And this is good, for it is the whole, not simply one part or another, that is always to be ministered to, nourished and lifted up in faith.

Before wading into the waters of risk and change, parish leadership bodies had to first develop and then operate out of a basic set of principles and/or premises. For the parish of St. Joseph, we began with only a few such operating principles, developing the rest as we moved along. And so, from our own experience we offer the following blueprint, in the hope that it will be of service to other parish communities who might find themselves looking at risk and change.

Operating Principles

- The good of the whole. We are a people who have a collective identity. Therefore, great care must be given to include all.
- Change simply for the sake of change is not very smart. In fact, it is dumb.
- Invite people to focus on a couple of key questions:
  - Who are we as church?
  - What is our purpose as church?
- Know that you will ruffle some feathers along the way. If any of us think that we can please all the people all the time, then we are in for a bumpy ride.
- There is an ancient principle about hunger that can be applied to spirituality: People go where the food is. Jesus himself knew this principle (cf. John 21:15-17). And so, when people gather at the Lord's table, FEED THEM.
- Educate along the way. We know that adult Catholics, for the most part, are not well schooled in their faith. As our retired bishop, Robert Rose, once said to me: "Educate, educate and keep on educating."
- Contrary to popular belief, Jesus' last seven words were NOT "We have always done it this way."
- It is God who is ever stretching us in faith, drawing us beyond ourselves.
- FAITH, not money, is the starting point. Still, there will be moments when it is necessary to write and/or speak about finances. Like any other enterprise, it costs to be church.

And in the same way, there is that great pearl of wisdom that serves to remind us that we get what we pay for.

- Look to hold up one or more persons who might serve as models for all of this. For the people of St. Joe's, we found ourselves looking to the example of St. Francis of Assisi....

.... Perhaps the most familiar of Saints throughout the world, one of the more striking characteristics of Francis was his capacity to risk and to dream BIG. The story goes that Francis, in the beginning of his conversion process, visited a small, abandoned and crumbling little church named San Damiano. While there, the crucifix which hung near where the altar once stood is said to have spoken to him "Go and rebuild my church." Francis worked tirelessly, even through the cold winter months. At first alone, he was later joined by a few companions. After much work and seemingly little progress, the crucifix again spoke to him "Go and rebuild my church." Then he got it. Francis was not being asked to rebuild a structure. Rather, he was being called upon to rebuild a people.

Nearly 800 years after his death, we still find ourselves much like Francis found himself before his great awakening. How much time, energy worry and resources do we continue to put forth in the building, maintaining and rebuilding of structures? In and of itself, we learn from Francis, this is not a bad thing, provided we are first concerning ourselves with the business of rebuilding God's people.

Nearly 800 years after his death, perhaps the greatest legacy of St. Francis was his capacity for taking risks and dreaming BIG, all the while growing to place his trust completely in God.

The remaining principles, from this blueprint or sorts, are of specific concern for pastors.

- Be strong in belief and conviction about what we are doing. If a pastor does not believe in what we are about, the parish will know it. Do not grow weak.
- We have been charged with being models and teachers of the faith, a charge that was given on the occasion of our ordination to the diaconate, at which time the bishop placed

the Book of the Gospels into our hands while speaking these words:

> RECEIVE THE GOSPEL OF CHRIST,
> WHOSE HERALD YOU NOW ARE.
> BELIEVE WHAT YOU READ,
> TEACH WHAT YOU BELIEVE,
> AND PRACTICE WHAT YOU TEACH.

-Rite of Ordination of a Deacon

- Be aware of and in tune with the themes of sacred scripture and the liturgical seasons. Make use of these themes, which are always intertwined with our current experience.
- There was only one Messiah, and it was not I.
- If we are to lead others in prayer, then we must be nurturing our own prayer life. In light of the preceding principles, I frequently find myself returning in prayer to the words of the late Theologian, Karl Rahner:

The priest is not an angel sent from heaven. He is a Man chosen from among men, a member of the Church, a Christian.

Remaining man and Christian, he begins to speak to you the word of God.

This word is not his own. No, he comes to you because God has told him to proclaim God's word.

Perhaps he has not entirely understood it himself. Perhaps he adulterates it. But he believes, and despite his fears he knows that he must communicate God's word to you.

For must not some one of us say something about God, about eternal life, about the majesty of grace in our sanctified being: must not some one of us speak of sin, the judgment and mercy of God?

So my dear friends, pray for him. Carry him so that he might be able to sustain others by bringing to them the mystery of God's love revealed in Christ Jesus.

- Return regularly and prayerfully to all of these operating principles.

Laying The Ground For Change

Having easily connected with the parish as a whole, and having come quickly to feel at home, I found it very easy to be up front and direct with the people of St. Joe's. For the most part, we were a good fit for one another. Also, I discovered that the parish's collective mind and heart were both open and flexible, which was a strong indication that the parish was primed for change. But without a doubt, I knew that I would find a few of those individuals, good people nonetheless, who balk at the very thought of change. Without their knowing it, the same individuals would stretch and challenge me along the way. They would be a visible reminder of that commandment which is sometimes the most difficult to keep: Love one another. And further, these individuals would serve only to strengthen my resolve to do that for which I have been ordained, namely, preach and teach.

With this said, how does one begin to broach the subject of risk and possible change? As previously stated, be up front and direct. This is a must. But equally important is the need to ground everything within the context of the parish, its own situation and its relationship to the larger church. In so doing, it becomes more likely that the parish, as a whole, will gladly take risks, daring to try things different from that with which they are familiar. And in the end, one need not be surprised if the parish comes to claim the process as their own.

How is it, then, that the people of St. Joe's came to take some risks and even try on change? On a particular weekend in the fall of 1997, my preaching focused around the following question: How many of us have a brother, sister, mother, father, son, daughter or friend who are minimally connected to the Catholic Church, or who no longer consider themselves to be "Catholic?" The response was the immediate raising of a hand by virtually every adult member of the congregation, including myself. Attention was then drawn to the gaping holes within our worship space, a vast amount of empty pews. Then, looking at our makeup along with the fact that the parish was continuing to decline, I began to wonder aloud: If we continue as we are, I am prepared to make a guarantee, which is something I do not often do. The guarantee is that if we continue as we are, here and at many of our parishes, we will see only more and more empty pews at Liturgy. And this will continue until such time that bishops find themselves arbitrarily closing parishes, which is already the experience in many dioceses. This was presented not as a threat, but a simple reality.

In that moment, I continued by reminding a people that as church we do not exist only for weekly Liturgy. Nor can we continue to exist as a parish simply because of our personal and emotional attachment to old buildings, etc., although these feelings and emotions must be honored as a significant part of ourselves as well as our faith journey. And finally, there was a statement of my personal belief that if there is life and promise within a particular faith community, then great effort should be given toward nurturing that life, that it might grow.

Where To Start

When entertaining thoughts of asking a parish, any parish, to take a critical look at themselves and consider taking some risks, especially when speaking about things as a matter of survival, there can be no better starting point than the Liturgy, which I often refer to as "the heart of it all." Before diving into the stuff of Liturgy, however, one must first tend to a number of preliminary issues in order to set the tone.

Preliminary Issues

- Perhaps some of the most familiar words from the Second Vatican Council (1962-1965) are those concerning the Liturgy. Within the council documents the bishops spoke eloquently of the Liturgy as "...the summit toward which the activity of the Church is directed" and "...the fount from which all the Church's power flows." (Vatican II Constitution On The Sacred Liturgy – chapter 1, #5)
- Make the distinction between "Mass" and "Liturgy." These two terms are often used synonymously, especially when referring to Catholic worship, yet they are very distinct in their meaning and significance.
  - The word "Mass" has its origin in the Latin "Missio" or "mission." "The Mass is ended, go in peace." Sound familiar? For about four hundred years, until the Second Vatican Council, "Mass" was the favored term for Catholic worship. It remains in common use to this very day, by priests and laity alike. However, the term "Mass," as well as the experience of Mass, has always been somewhat constricted as a result of the following:
    - ➤ Throughout the Latin era the Mass was highly clericalized, as if worship were a commodity or property belonging only to the priests.
    - ➤ On the part of the individual Catholic, worship had become very privatized. This is seen no more clearly than in this statement, which I continue to hear occasionally: "I / we have got to go and hear Mass."
    - ➤ In a nutshell, the result was that Mass was a private worship experience, which focused on priest-centered activity, with the laity assuming the role of passive observer. Shut out, in a sense, is it any wonder why other practices crept into Catholic worship? Practices such as praying the rosary or a novena, privately, during the Mass?
  - With the Second Vatican Council there was a return to the use of the more ancient term for Catholic worship,

namely, LITURGY. Greek in its origin, the word is defined as "work of the people." As such, the very shape and character of Liturgy are rather distinct from the experience of "going to Mass." Liturgy would have us approach worship in this way:

➤ By its very definition, Liturgy belongs to the entire faith community. This is not to diminish priesthood in any way. Without a doubt, the presiding priest has a unique role to play. But no longer is it simply "Father's Mass." Liturgy is our act of public worship, and everyone who gathers within the worship space is called upon to participate in the sacred action of that Liturgy, echoing the Council's call for the full, conscious and active participation of the faithful in Liturgy.

➤ Where "Mass" was a room full of individuals who were privately "going to church," Liturgy is to be the church in all its members (one body) coming together for a communal act of worship.

This distinction between "Mass" and "Liturgy" might seem trivial, but the differences are huge. And it is around this distinction that I have come to believe that there are fundamentally two groups of people within most Catholic congregations today: Those who "go to Mass" and those who gather together to celebrate **their** Liturgy. There is a big difference between "going to church" and "**being** church." One has to do with personal obligation at a private level; the other has to do with **US** at a collective level. Liturgy reminds us that we are church. And we are church together, not simply a group of individuals who "go to church."

Having spoken of the distinction between "Mass" and "Liturgy," it is important to invite a congregation to reflect on a couple of related questions:

➤ What is YOUR experience of Catholic worship?

➤ Do you simply go to Mass? Or do you gather with others, as church, to celebrate Liturgy?

Around this distinction, it is equally important to invite priests to reflect on similar questions:

> ➤ What is YOUR experience of Catholic worship?
> ➤ Do you simply "say Mass?" Or do you **preside** at Liturgy?

About priests and bishops, I am going to tip my hand and make this general statement: It is my belief that the majority of these men continue to "say Mass." They do not preside at Liturgy.

- It is necessary to briefly address the issue of obligation, specifically the obligation to attend Sunday Liturgy.

  The obligation to participate in Sunday Liturgy is rooted in the sacrament of Baptism. In the sacrament one is baptized not only into the very life of God, but into the life of a people as well, a people called "church." In Baptism God has said "Yes" to us. And in turn, we have said "Yes" to being God's people. The obligation to attend Sunday Liturgy is thus grounded, from apostolic times, in the very nature of our being church, a term which comes from the Greek **ekklesia**, meaning **assembly**. As church, the Sunday assembly remains the foremost sign and manifestation of what it means to be church. We are a people who gather as one, that we may be transformed into the Body of Christ.

  In many respects, Catholics have all but lost this ancient sense of our obligation to be church. This is reflected in the statistics that, for years, have shown that less than 50% of adult Catholics regularly attend Sunday Liturgy. Some would suggest that this is a result of contemporary society's great emphasis on individualism. Others, including some priests, simply point fingers at inactive or "fallen-away Catholics" (an expression which I personally despise) as if to imply that these folks are no longer people of faith. I, however, firmly believe that there are two primary reasons why large numbers of Catholics are not regularly participating in Liturgy:

o A similar problem emerged centuries ago, to which church leaders responded by not only reminding people of their baptismal obligation to be church, but then went on to say that failure to attend Sunday Liturgy is a mortal sin. Much of this was the result of some very poor theology and equally poor piety that had crept into the church, with a near-total emphasis on the sinfulness and unworthiness of the individual. In effect, church leaders threatened the people of God with Hell tomorrow, if they didn't go to church today. Sadly, some of this attitude is alive and well in our own time, especially within the hearts of many of our present church leaders.

o We are not feeding the people when they **are** at Liturgy. This is what I consistently hear from those who are no longer attending Liturgy regularly, or who have left Catholicism altogether.

In my mind, and based on experience as well, the solution to this is very simple: Keep the good theology, and dump the garbage. Return to Jesus' command, "Feed my sheep." If we are doing this, and doing it well, it will not be necessary to harp about people's obligation to be church. They will BE church. So, let's stop beating and threatening God's people. Instead, let's return to imitating Jesus' example of loving people into the Kingdom.

• It is important to emphasize that trying something different does not necessarily mean something "new." As is often the case among Catholics, what we like or dislike about church is colored, shaped and very much limited by our lived experience. We often fail to remember that the church has a history and lived tradition that long precedes any one or all of us who are church today.

And so it was, at St. Joe's, that as we began to do things differently, I reminded the parish that we were doing nothing "new." Rather, we were looking backward to the old, hoping to discover something of great value. And in our search, we were not disappointed.

- It will also be helpful to speak about this wonderful word, "tradition." Derived from the Latin "traditio," it means "to hand-on to" or "the handing-on to" another.
  What, then, is the "tradition" or that which is being handed-on? The very core of our faith, namely, the faith of the apostles.

Within church circles, this word "tradition" is often misused as a result of having been associated with practices that should properly be called "custom(s)." As such, one might find it helpful to differentiate between Tradition (with a capital "T") and tradition (with a lower-case "t").

- Speak to and, more importantly, DO ritual. From the beginning it seems that humans have been created as ritual beings. As such, we have developed various rituals, actions that may or may not be accompanied by words. And these ritual actions serve to mark or remember significant moments and/or events in the life of an individual, a group or persons, or perhaps an entire culture. At the spiritual level as well, ritual action serves to give expression to the deepest of truth or beliefs for which no amount of words are ever quite adequate.

  Catholic Liturgy is itself a ritual, within which there are a number of ritual actions on the part of the presiding priest as well as the people gathered. Unfortunately, many of our parishes (and thus, many Catholics) do not encounter and experience good ritual activity. And so, to presiders, musicians and Liturgy planners, I offer these few words: Do the ritual, and do it well. But do it in this order: First, give people the experience. Then, if necessary, explain and educate about the particular ritual. This approach is much in line with St. Anselm's statement regarding catechesis, or instruction in the faith, in which he said "Faith is first caught, then taught." Toward this end, I offer a concrete example:

  One of the very first things we did differently, at St. Joe's, was to return to the use of incense at both weekend

Liturgies. All that was said in advance was that we were returning to this practice, which in recent years has been relegated to funeral liturgies. And even there, the ritual is often done poorly. Aside from only two or three complaints about the smell, the parish quickly came to appreciate the ritual (the experience). Several weeks later, there came a brief catechesis (the instruction), the heart of which the congregation had already grasped:

o Liturgy should effectively touch all of our human senses.
o By involving our sense of smell, the use of incense assists in drawing a people all the more wholly and deeply into the Liturgy.

Historical reasons for the use of incense:

o Symbolic of the Israelites' journey through the desert.
o Sign of the raising-up of our prayers and the lifting-up of our very selves to God, that we might be a pleasing fragrance to God.
o Sign of reverence. As our ancient Jewish and Christian ancestors incensed that which was held to be holy, so we do the same when we incense the holy word, the sacred altar, bread and wine, and a people who are called to continued growth in holiness.

With the groundwork laid, and having addressed some general principles and preliminary concerns, we now move into the Liturgy, the heart of it all.

Liturgy

Prior to doing a formal instruction on Liturgy, it was my intention to first give people something of the experience. This was done by way of preaching. On a particular weekend I delivered what I have referred to as a "song and dance homily" about Liturgy and all that goes into it. This proved helpful in striking a chord, setting the tone and even whetting a people's appetite. But because our story is now being told in written form, for obvious reasons it is impossible to include the song and dance here. What follows, however, are the two educational pieces that were presented to the people of St. Joe's.

Eucharist at a Glance – The Short Form

The proper theology behind the Sunday Liturgy is bound up in the Eucharistic Assembly. To be a Christian means to be a member of the Christian assembly, the church realized in the local community of the parish. As demonstrated repeatedly throughout salvation history, continuing today and into the future: God calls each of us by name, but saves us as a people. Today's unhealthy individualism in American society and elsewhere, which colors religious attitudes, is in direct antithesis to the one Body of Christ consisting of interdependent members who need one another.

Stated in words attributed to St. Augustine: "The Church makes the Eucharist and the Eucharist makes the Church." Its meaning? In celebrating the Eucharist we express our truest identity as a Christian people; the celebration of God's Word and Sacrament in turn forms us more completely into the Body of Christ. Our very unity, then, as a church is Eucharistic, that is, brought about and sustained by the Word of God and the Sacrament of the Lord's Body and Blood.

Definitions

- Liturgy – work of the people.
- Eucharist – Thanksgiving, receiving the Body and Blood of Christ, the sharing of a meal which celebrates God's love for us **AND** our love for one another

In the Eucharistic Liturgy we celebrate the Lord's presence

- In the people who have gathered in His name,
- In His Word,
- And in Communion.

Themes of Eucharist

Thanksgiving...Meal...Welcome...Family...Celebration... Sacrifice...
Identity (as a community, as a people)...Death and Resurrection (repeating, prolonging, renewing, continuing, making present)... Salvation...Unity...Worship...Praise...Prayer...New Life...
New Covenant...Gift...Peace...Growth...Service (or being sent)...

Outline of the Eucharistic Liturgy

* Gathering and preparing,
* Liturgy of the Word,
* Prayers of the Faithful (a bridge),
* Preparing of Table and Gifts,
* Liturgy of the Eucharist (Eucharistic Prayer and Communion Rite),
* Concluding Rite (Mission...being sent forth).

Eucharist – The Long Form

The following is an expanded outline or walk-through of the Liturgy that, for Catholic Christians, has drawn us together for nearly two thousand years. Keep in mind that Liturgy (the work of the people) does demand a great deal of work.

Gathering

When does the Liturgy begin? Formally, it begins with our gathering song, most often a hymn of praise and joy. Standing, we raise our hearts and voices in song as a sign of reverence and welcome of the One who has called us together.

Procession

Relative to our gathering is the procession itself (servers, liturgical ministers and presider), which reminds us that it is Jesus

Christ whom we reverence and welcome in our midst…God present among us in the people gathered, in the spoken Word and in the Eucharist.

Vestments

The liturgical vestments, worn by the presider, are themselves reminders of the above. The large, flowing chasuble is also a visible sign of the wideness of God's love and mercy. Often bright in color and differing according to the liturgical seasons and feast days, the priest's vestments further remind us that there is a festive nature to Liturgy. Worship, therefore, is a celebration…a party. God is present among us. Now that's worth celebrating!

Informally, Liturgy begins long before the gathering hymn. It begins with readying ourselves at home, pre-disposing ourselves in heart and mind, putting ourselves into a mode of anticipation and receptivity. We are not "going to church." Rather, we are going to **BE** church. Assembled as church, we will welcome the Lord Jesus Christ in all who are gathered; we will reverence Him in Sacred Scripture; we will receive Him in the Eucharist (a most intimate encounter).

A wonderful image of Eucharist is found in the very act of our gathering, as the doors open and all the members of Christ's body assemble. Do you want to know what Christ looks like? Just listen and look about as the Body of Christ comes together and takes shape for Liturgy. What a beautiful picture! We have life, color, diversity, wealth, poverty, etc. And each member of that body brings their entire self to Liturgy (their joys and hopes, their griefs and anxieties), laying it before God with the hope, trust and expectation that God will act, bringing us closer to completeness and wholeness in the Lord.

Our gathering evokes yet another image, that of sacred space. To walk through these doors is to enter a different reality. In many of the Eastern Rites the church building or worship space is referred to as "the place where heaven and earth come together." Hence, our church building is no "ordinary" meeting place.

Water

Signing oneself upon entering the church is a reminder of our baptism: In Christ we have become a new creation. We gather that we may continue to grow and develop as that new creation.

Altar

On behalf of the community, the presider reverences the altar with a kiss. The altar (table) is a symbol of Christ. (Remember that a symbol not only points to another reality, but also participates in the very reality that it signifies) It is upon Christ that we place our very lives.

Penitential Rite

Here, we take a moment for self-evaluation. In effect, we ask ourselves "How well, how faithfully have we (as individuals and as a people) lived as disciples of Christ since our last gathering?" In varying forms, we then ask the Lord for mercy, forgiveness, healing and peace.

Gloria

The Gloria is a hymn, the hymn of the angels. Liturgical experts have insisted for years that as a hymn, the Gloria should be sung. AND if it is not going to be sung, then the Gloria should be omitted. One way to illustrate this point is to use an example, that of the song HAPPY BIRTHDAY. When those words are spoken or recited (try it), the effect is obvious. The same holds true with the hymn of the angels. Still, many a pastor insists on reciting the Gloria at Liturgy. And just like reciting the words to Happy Birthday, the effect is painfully obvious.

Liturgy of the Word

Having gathered and recollected ourselves, we now break open the scriptures...that the Lord may speak to God's people.

Liturgy taps (or should tap) all our senses. Hearing is no exception. From the earliest days it was understood that when the people of God gathered for Liturgy, God's Word was to be SPOKEN and HEARD. The Word of God is to be presented as a proclamation of GOOD NEWS, to be heard by the people who have gathered to hear it. Further, our eyes should be fixed on the person who is proclaiming, for at that moment he or she is the voice or mouthpiece of God.

The scriptures proclaimed at Liturgy are stories of God, stories of a people and stories of faith. The stories are retold time and time again, that we might find ourselves within them. Typically, the Sunday readings are arranged as such:

- Old Testament reading
- Psalm (which should be sung because the psalms were written as hymns). As a response, we are claiming God's Word as our own.
- New Testament reading
- Gospel: We stand and focus all of our attention on the gospel and its proclaimer, this as a sign of our utmost reverence and attentiveness to Jesus' own words.

Homily

This is a time for the presider (or another who may be preaching) to further break open and expand the scriptures that were just proclaimed. The homily is about connecting, **our** connecting with God's Word, that it might become within each of us a LIVING WORD.

Having heard and reflected on God's Word, we now begin moving into the Liturgy of the Eucharist. This is a transition moment, a bridge of sorts, marked by the Creed and the Prayers of the Faithful.

Creed

United, and with one great voice, the community makes a public profession of its faith. The Creed is both our response to God's Word **AND** a re-commitment of ourselves to living out the faith we profess.

Prayers of the Faithful

The Prayers of the Faithful (or Intercessory Prayers, as they are sometimes referred to) are the means by which the community publicly entrusts to God all their needs as they make their journey. God already knows our needs. But the benefit of asking, of literally giving a voice to our needs, is for ourselves, so that we come to know (as God already knows) what we truly need in the life of faith. These prayers are always presented for the people gathered, for the universal church and for the world. By responding "Lord, hear our prayer," we are making the petition our own prayer.

Liturgy of the Eucharist

Preparing the Table

Accompanied by song or an instrumental piece, the altar is readied for Eucharist. During this time we take up our collection, our tithe, a sign of the giving of ourselves to God as faithful stewards of all that God has given to us.

We also bring forth the simple gifts of bread and wine as a symbolic offering of our very selves to God and to the service of God. In ancient imagery, this is referred to as "the sacred exchange:" In the presentation of the gifts of bread and wine we are giving ourselves to God; God then changes the gifts, making them holy, and gives them back to us as Jesus Christ.

Prayer over the Gifts / Preface Dialogue / Sanctus

With the altar prepared and now focusing on the table of Eucharist, the community lifts their hearts and lives to the One who comes in the name of the Lord.

In Catholic worship we tend to not do this very well. Our movements and gestures often do not correspond. For example: We pray "May the Lord receive this sacrifice/this offering from our hands…" Then, everyone immediately stands up when, in fact, they should remain seated until the presider invites and gestures the congregation to stand. And this occurs during the preface dialogue (unless incense is being used), which follows the prayer over the gifts. Hence, the preface dialogue:

Presider:	The Lord be with you.
People:	And also with you.
Presider: *	Lift up your hearts.
People:	We lift them up to the Lord.

- This is the moment to stand. Again, if incense is being used, the congregation will have been invited to stand prior to the prayer over the gifts as well as the preface dialogue)
- Interestingly, we should remain standing from this moment until we return to our seat after receiving Eucharist (communion). With the reform and restoration of the Liturgy at Vatican II, the General Instruction of The Roman Missal states: "…the people should stand from the prayer over the gifts to the end of Mass…" Only in very recent years have some parishes begun to implement this practice. Sadly, however, the current leadership in Rome is preparing to issue (and enforce) new liturgical directives that will have us return to the pre-Vatican II posture of kneeling throughout the Eucharistic prayer. These directives are due out at the end of the year 2004.

Eucharistic Prayer

In its varying forms the Eucharistic Prayer is quite simply a re-telling of the story of salvation history, of which each one of us are a part. In a sense, it is also an expanded version of the Creed.

The Eucharistic Prayer contains a number of key elements:

- Praise to God.
- Thanksgiving for all that God has done and continues to do.
- Epiclesis: Calling upon and invoking God's Spirit to make us, and our gifts, holy.
- Institution Narrative and Consecration, the words and action of Christ.
- Anamnesis, or remembering. In the very act of remembering, the event is present before us.
- Intercessions: Praying with and for all people of all times and places.
- Doxology: Our prayer of praise, to which all respond 'AMEN.'

The Lord's Prayer

Prayer for all that we need: Daily food, Bread from heaven, food for the journey (Viaticum), holy gift for a holy people.

Rite of Peace

This ritual sign, as the community prepares to receive the Eucharist, is a prayer for God's peace to find a home within each one of us, within the church and throughout the world. As we come to the table, we pray as Jesus prayed: "That all may be one."

Breaking of Bread

- A visible reminder that Christ was broken for us. As He was broken, so must we be willing to be broken for others.

- The <u>Lamb of God</u> is sung during the breaking of the bread, further reminding us that it is God who takes away our sins, makes us holy and gives us new life.
- Fractioning Rite: A small piece of the consecrated host is dropped into the chalice as a sign of union, the coming together of the human and the divine.

<u>Communion</u>

This most intimate moment is our feeding on the Bread of Life, Jesus Christ. The sacred exchange is now complete.

From personal experience I have observed that the communion rite, perhaps the most intimate moment within the Liturgy, has become somewhat hollow. This is due, in part, to the fact that when anything becomes routine, its depth and significance are at risk of being removed from one's consciousness. The rite is further robbed of its richness by, sadly enough, priests who themselves have allowed Liturgy and perhaps even priesthood itself to become far too routine. This was clearly the experience at one parish with which I have had some familiarity:

The pastor was something of a businessman. A well-organized person, he possessed a real talent for raising money. And he was, as the saying goes among priests, a "brick and mortar guy." But he was not, in my opinion, a liturgist. At one point I heard him utter these words concerning Liturgy: "You know my philosophy: Get 'em in and get 'em out!" In my mind, the only appropriate response to such a person/priest is "So much for feeding God's people."

We must, therefore, rediscover the depth and significance of the communion rite. When we come forward to commune, to receive Jesus Christ in the Eucharist, it may remain a brief moment. But it is to be a profound experience, for it is an encounter with the living God. At that moment, there is made a faith statement by the Eucharistic Minister. The recipient or communicant then makes a faith response. In black and white it looks like this:

- The Eucharistic Minister makes this faith statement: 'THE BODY OF CHRIST' (or, when receiving from the cup, 'THE BLOOD OF CHRIST').
- The communicant responds with a faith statement: 'AMEN'...meaning YES, SO IT IS, I BELIEVE, YES, WE ARE.

Prayer after Communion

Words of gratitude: For this great gift of God, and for our encounter **with** God.

Rite of Sending Forth (Mission)

Having heard God's Word and having been fed on the Bread of Life, Jesus Christ, the community is now sent forth to **BE** Word and Eucharist for all others.

Hymn / Song of Praise

We should do nothing less than sing out in joy and gratitude for what God has done in our midst. As a priest, I take great care to never say the words "The Liturgy (or Mass) is ended...." There is no end to Liturgy. So, keep the song alive until we gather again with our God in prayer.

After all that has just been said about Liturgy, this is a good point at which to pause and address a topic that is related to Eucharist. It was in 1999 that a small but vocal group of Catholics, primarily from two neighboring parishes, put forth the word that they were looking to establish a small chapel in which they could return to an old practice or custom known as Perpetual Adoration. Among the ranks of those who were promoting this practice were the respective pastors of the two neighboring parishes. And their support of Perpetual Adoration came as no surprise to me whatsoever. All of this, however, came about as the people of St. Joseph found themselves in the midst of being educated, stretched and opened to a larger vision of themselves, the church, the Eucharist, etc. Because a small number of my own parishioners asked for my thoughts on Perpetual Adoration, I put together the following piece that went out to the entire congregation. And I did so with the knowledge that it would also find its way into the larger Catholic community. Hence, it would be one method by which I could invite still others to stretch, to think a bit larger about their faith, about Eucharist, and perhaps even their very selves. Thus I am pleased to include, here, some reflections on Perpetual Adoration (sometimes referred to as Eucharistic Adoration).

19 July 1999

Perpetual Adoration

Personal Thoughts
Fr. David Harpe

I am a person who believes, quite strongly, that we should have a good understanding of why we do the things we do. As a priest and person of faith, the same must be said of our religious practices. In the form of a question, one might ask "What is the faith and theology which underlies our system of religious beliefs and practices, either in part or in its entirety?" And if the theology is weak and lacking, or inconsistent with sacred scripture, then perhaps that particular practice should rightfully be shelved.

In recent months there has been a growing interest, locally and nationally, in reviving the practice of Perpetual Adoration of the Eucharist. Quite frankly, I personally do not support a return to this practice or devotion because it does not appear to be in concert with the theology of Eucharist that is presented in the New Testament, the same theology which has been handed on to you and me. Pardon me for speaking bluntly, but it seems to me that the only thing being perpetuated by perpetual adoration is poor theology. Consider the following:

By the eleventh century, abuse and corruption were widespread throughout the church. Some of the more notable forms of abuse were:

- Civil authorities and church leaders in cahoots with one another, motivated by greed for more wealth, land, possessions, etc.
- Clerical immorality, a result of men seeking priesthood in order to attain wealth, prestige and the like.
- The clericalization of worship, by which the liturgy came to be viewed as property of the clergy. Screens were erected in worship spaces, literally separating the clergy from the laity. God's people had become passive observers to the sacred action of liturgy.
- Box Masses, the factory-like operation of priests lining up to say or sing a private Mass for a particular intention. This was fueled by superstition on the part of the laity. For example: If I buy a Mass, then something good will happen to me. Or, if I pay more for a sung Mass, God's favor to me will be even greater. And this practice was further fueled by greed on the part of the priests who carved out a pretty good living by saying Mass. Directly stated, the more Masses said, the more money.
- And of course, there was the rampant selling of indulgences.

All things considered, the eleventh-century European church was readily plagued with corruption. Within that context the soil had become ripe for still further questionable activities and practices, effecting a shift even in our understanding and theology of Eucharist. Prior to the year 1000, Eucharist was understood and celebrated as a sharing in the table of the Lord, to which all were welcome. After the year 1000, we saw this major shift regarding the Eucharist:

- The primary concern was to make Christ present
- The move of the laity away from active participants to passive observers of Liturgy. Feelings of awe and reverence emerge as the appropriate posture for liturgy. And with this comes a growing sense of unworthiness. This feeling of unworthiness became so extreme that large numbers of people stopped receiving Christ in the Eucharist. By the way, this is part

of the framework which ultimately led to the "obligation to attend Mass" under the threat of sin, something that haunts us to this day.

- Privatization of worship: a "me and God" mentality. No longer was liturgy experienced as a community's celebration of its life in Christ, but rather as a gathering of individuals for the purpose of praying quietly and reverently watching the priest make Christ appear.
- Perhaps the most radical shift in Eucharistic theology would come when people began viewing the consecrated host as a relic. Emphasis was heavily placed on seeing the host, for to see the host was to see Jesus. And this paved the way for other practices and, in some instances, very strange claims around the Eucharist:
  - o Placing the host in a large reliquary or monstrance and leaving it on display for public viewing. In time these reliquaries would develop or evolve into elaborate and ornate objects, accented by gold and precious jewels..... the very trappings with which Jesus did not wish to identify himself or his kingship.
  - o As a relic, the host was viewed as the most important symbol of power. Instead of receiving and **eating** Eucharist, people were interested in **looking** at the host in the hope of gaining power. Rather than bringing Eucharist to the sick and the dying as food for their journey, people were simply waving the Eucharist over the sick and dying.
  - o Emergence of the belief that if one were in the presence of the consecrated host, one would not age.
  - o Claims to have seen bleeding hosts.

This sounds about as close to superstition and magic as one can get without calling it such. And to reduce Eucharist to an object for viewing and adoration seems all but totally divorced from the New Testament and its sense of Eucharist.

Now, let's move to the present.

In the 1960's the Second Vatican Council called the church to a profound reawakening of faith. While the council mandated a number of reforms as part of this reawakening, none have impacted us more than the reforms and restoration of the sacred liturgy. Here is a thumbnail sketch of the reform:

- The liturgy is the work and action of the community of believers, not the property of the clergy. All have a role to play
- Liturgy, by its very nature, is communal. It is not private (me and God).
- The obvious physical changes were implemented to assist the community in reclaiming, participating and engaging in the sacred action of the liturgy.

At the very heart of liturgy is Eucharist. And it is the theology of Eucharist, as rooted in scripture and the early years of the church, which I believe many people have failed to grasp. Instead, many continue to come to liturgy with a very narrow, private and personal focus on Eucharist. Unable to connect with the Eucharistic liturgy and its broadness, they find themselves frustrated, on the edge and minimally involved. This is always the case when one struggles to make that proverbial round peg fit into a square hole.

To those who find themselves disconnected and on the edge, to all who read and reflect on this commentary, Eucharist is not an object to be locked up, viewed and guarded. It is much bigger than that, indeed much larger than any one of us or all of us together. Consider some of the many meanings and themes of Eucharist:

- Thanksgiving
- Welcome
- Celebration
- Sacrifice
- Death & Resurrection
- Bread of Life
- Salvation
- New Life
- Gift
- Growth
- Presence

- Meal
- Family
- Community
- Unity
- Worship
- Praise
- Prayer
- Peace
- New Covenant
- Being Sent ....To Serve

Perhaps the most endearing image and meaning of Eucharist for myself is that of bread. If bread is to nourish and sustain life, then it must be eaten. So it is with Eucharist: It is to be broken, taken into ourselves so that it may grow in us, that we may grow into Eucharist. And what we have received as a gift we must then give as a gift....by living and being Eucharist for others. **This** is perpetual adoration.

Response / Reaction to Risk and Change

As we began taking risks and implementing change at St. Joe's, there were some who found it difficult because, in their own words, "it was different." But they remained nonetheless, faithfully supporting and participating in the life of the parish. For a few individuals it was simply too difficult, or rather too different, and they ultimately decided to leave the parish. But for the large majority of our parishioners, we apparently struck a beautiful chord. And this chord would resound far beyond the walls of St. Joseph, even beyond Muskegon.

There is no better way of describing the parish's positive response and overall acceptance of change than by simply highlighting what were the key elements / experiences of the community, for it was truly **our** experience.

- BREAD – In 1999 we were informed that the religious order of sisters who supplied many parishes with communion wafers would no longer be doing so. Instead of looking to another religious order, or turning to a religious goods distributor, we at St. Joseph returned to the older tradition of using unleavened bread. Also referred to as "substantial bread," it has more substance and ritual value than the little wafers that we have grown accustomed to in Catholic worship. Again, we did a little instruction on the use of bread at liturgy: The simple gifts of bread and wine represent our own lives. Given to God in the liturgy, God then gives them back to us as Jesus Christ, the Bread of Life.

We then invited individuals and / or families to consider taking turns baking the bread that would then become our Eucharist (communion). The response was overwhelming. Young and old alike, especially our children, found great delight in serving the faith community in this way, in this ministry.

Despite the positive response, there were a few individuals who expressed some concern with regard to practices of piety or, quite frankly, their blatant dislike for anything "different." With that, let me share a bulletin column in which I responded to these individuals:

Harpe Melodies

9/5/99

I do not wish to labor over the issue, but after using the unleavened bread at last week's liturgies it seems clear that a few more words are called for....

From the rumor mill or through the grapevine, it was reported that upon entering the church and seeing the loaves of bread one individual made the remark "Another one of his ideas." (I believe that person was referring to me.) I would like to take credit for this one, but I cannot. Credit for this belongs to Jesus. It was and is.... another one of HIS ideas.

On the practical side, a couple of individuals found that it is nearly impossible to receive on the tongue when using unleavened bread instead of hosts. Perhaps the greatest risk is that the Eucharist might fall to the floor when attempting to place it on another's tongue. In one instant, last weekend, that is precisely what happened. Much to my surprise, some still raised the issue of sinfulness and unworthiness as the reason for not receiving Eucharist in their hands (e.g., "I can't touch the host because I am unworthy"). To this I often respond "Then perhaps we shouldn't receive at all, because we are still touching the host." Recently, I heard a wonderful story about this very issue....

....It was several years ago, when we began returning to the older practice of taking and receiving Eucharist (in our hands). I am told that a local pastor was approached by a small yet vocal group of parishioners who complained about receiving Eucharist in the hand. They told the pastor they could not do this because of their unworthiness. Further, they asked the pastor to consider the hands, and how much sinfulness in committed by our hands. The pastor responded by asking this group if the tongue makes us any more worthy to receive Eucharist. He then asked them to consider how much sinfulness is committed by our tongues, how much harm is done by the words that come from our mouths.

I hope this story causes us to think and reflect on our attitudes about Eucharist. By the way, I nearly forgot to tell you that that pastor's name was (and is) Robert J. Rose, our current Bishop.

- INCENSE – As previously stated, the use of incense was a welcome addition to the liturgical celebrations at St. Joseph.
- NO MISSALETTES – Instruction was given about God's Word being proclaimed and heard at Liturgy, not read by the congregation. Attention was drawn to the ambo, the table from which the word is proclaimed and upon which that word is enthroned. While the instruction and the change (the doing away with the missalettes) were well received, there were still those few persons who found it difficult. Their chief complaint was of a practical nature: "What if we are hard of hearing?" Not a problem. People were invited/challenged to draw closer to the word. And if in time they still found

it necessary to have the printed text before them, this was acceptable. But, the parish would not supply the missalette, except in cases of hardship. This became an occasion for encouraging these folks to take responsibility for their own issue. If they were adamant about having the written text in their hands, then we simply provided the address to which they could write (or call) and order their own missal. Then it would be their responsibility to bring the missal with them to weekly Liturgy.

The instruction on God's Word, mentioned above, appears here, in its entirety.

Harpe Melodies

11/16/98

GOD'S WORD PROCLAIMED

Over the past year there has been many occasions when I have encouraged the congregation to focus on **listening to** and **hearing** God's Word when it is proclaimed at Sunday Liturgy. With the recent approval of the new Lectionary (book of readings) which we will begin using on the first Sunday of Advent, it would seem that this is the appropriate time to do away with our use of the seasonal missalettes. Rather than simply "dump" the missalettes, I would like to take this opportunity to provide some catechetical and practical instruction on the matter.

Long before the scriptures appeared in written form, the stories of faith were handed on by way of an oral tradition. This is particularly true of the First (or "Old") Testament: Families and communities of believers handed on their faith by gathering together and telling the stories…again, and again, and again. Only with the passing of time, over many generations, did the stories come to be recorded in written form in order to preserve them. Preserve them from what? The growing fear that the stories could be lost or changed. And this fear was rooted in the very real experience that anything committed strictly to memory can indeed be permanently altered or forgotten altogether.

While the stories of faith have been recorded in written form for many centuries, Catholicism has maintained the ancient practice of proclaiming God's Word by way of story telling, when the community of believers gather for the sacred Liturgy. There was, however, roughly a 400-year period when this practice was all but lost. And this was during the Tridentine era, which lasted from the mid-16th century Counter-Reformation to the reforms mandated by the Second Vatican Council (1962-1965).

Nowhere would Catholics experience Vatican Council II more tangibly than in the reform and restoration of the sacred Liturgy. One of the great challenges of the Council was to reawaken our sense of the importance of God's Word in our lives, especially as it is proclaimed and heard within the Liturgy. And so began the move to return to a Liturgy in which the ancient practice of story telling is the method by which God's Word is loudly and boldly proclaimed. But because we were emerging from the Tridentine era, during which Catholics relied on an English translation of the Missal so as to know what was going on at Liturgy, the liturgical reform of the 1960's gave rise to the use of missalettes. The missalette was intended for use as a temporary tool of transition, a means for assisting us in returning to the older practice of worshiping in the vernacular (the language of the people). Thirty years later, however, it seems that this "temporary tool" has become something of a permanent crutch. The result is that when we gather for Liturgy we have a group of individuals who are privately and quietly reading the scriptures. And this is in opposition to the ancient sense of the liturgical assembly as a people, with a communal identity, who are gathered as LISTENERS, waiting to HEAR God's Word (Good News) proclaimed.

In the new Lectionary's <u>GENERAL PRINCIPLES FOR THE LITURGICAL CELEBRATION OF THE WORD OF GOD</u> there are numerous references that support the ancient practice of proclaiming and hearing God's Word when the church is gathered for the sacred Liturgy. The following are but only a few citations from the lectionary's general principles:

o When God communicates his word, he expects a response, one, that is, of listening and adoring "in Spirit and in truth" (John 4:23). The Holy Spirit makes that response effective, so that what is heard in the celebration of the Liturgy may be carried out in a way of life: "Be doers of the word and not hearers only" (James 1:22) GP – paragraph 6

o Accordingly, the participation of the faithful in the Liturgy increases to the degree that, as they listen to the word of God proclaimed in the Liturgy, they strive harder to commit themselves to the Word of God incarnate in Christ. GP – paragraph 6

o In the hearing of God's word the church is built up and grows. GP – paragraph 7

o Baptism and Confirmation in the Spirit have made all Christ's faithful into messengers of God's word because of the grace of hearing they have received. GP – paragraph 7

o Accordingly, the faithful listen to God's word and meditate on it... GP – paragraph 8

o A speaking style on the part of the readers that is audible, clear and intelligent is the first means of transmitting the word of God properly to the congregation. GP – paragraph 14

o There must be a place in the church that is somewhat elevated, fixed and of a suitable design and nobility. It should reflect the dignity of God's word and be a clear reminder to the people that in the Mass the table of God's word and of Christ's body is placed before them. The place for the readings must also truly help the people's listening and attention during the liturgy of the word. Great pains must therefore be taken, in keeping with the design of each church, over the harmonious and close relationship of the ambo with the altar. GP – paragraph 32

o In order that the ambo may properly serve its liturgical purpose, it is to be rather large, since on occasion several ministers must use it at the same time. Provision must also be made for the readers to have enough light to read the text and, as required, to have modern sound equipment enabling the faithful to hear them without difficulty. GP – paragraph 34

o   Along with the ministers, the actions, the allocated places, and other elements, the books containing the readings of the word of God remind the hearers of the presence of God speaking to his people. Since in liturgical celebrations the books too serve as signs and symbols of the higher realities, care must be taken to ensure that they truly are worthy, dignified, and beautiful. GP – paragraph 35

So ends the catechesis. Now for some practical suggestions, that we may better hear the word of God proclaimed at Liturgy:

o   Consider drawing close to the word of God (after all, it is Jesus Christ who invites us to come to the table of Word and Eucharist). Believe me, there is plenty of available space at the front of the church. For those who sit further back, the ability to hear God's word is greatly diminished.

o   We have begun printing the daily readings in the bulletin. Consider looking at the scripture readings for the coming Sunday, prayerfully reading them at home as a way of preparing for Sunday Liturgy. This is also a good way to invite many Catholics to open the Bibles that normally sit on our coffee tables or bookshelves.

o   You may wish to subscribe to a publication such as SHARE THE WORD, which is a good way for looking at and reflecting on the scriptures in anticipation of the coming Sunday.

o   If you absolutely must have the readings in front of you at Liturgy, then perhaps you could purchase a St. Joseph Missal and bring it with you to the Liturgy each week.

I hope this is helpful. May all of us be good proclaimers, listeners and witnesses to God's Word alive among us. Peace and blessings always, Fr. Dave

- CHILDREN GATHERED AT THE TABLE – Let's face it, the Liturgy is very adult in its shape. But our children are part of the congregation, and they are members of the church. How, then, might we help our little ones to better connect to an adult liturgy? Many parishes provide a children's liturgy of the word, which is a good way for them to break-open and reflect on God's word apart from all the adults. But can we not do more? I believe we can. And we must.

  One of the very first things we began doing differently at St. Joe's was to invite our little ones to gather around the altar (table) during the liturgy of the Eucharist. Once the table was prepared and the incensing of all was complete, our children were invited to draw near. And did they draw near! There they would remain until returning to their families during the rite / exchange of peace.

  By and large, this was in incredible experience for all. Initially, however, there were those few adults who grumbled. Their greatest concern was that the children might be something of a distraction. How might I respond most effectively to this concern? The opportunity presented itself at a particular weekend liturgy in which children were very much a part of the Gospel reading.

  In the course of the homily, I drew attention to an issue that has plagued the church for many years: catechizing large numbers of children only to have many of them not remain connected to the church. In short, we instruct our members while they are children, which was not the example of Jesus who preferred to instruct adults. For the most part, we give children the basics of our faith, preparing them for sacramental celebrations such as First Eucharist, Reconciliation and Confirmation. Not that this is wrong, but catechesis and ongoing formation in faith tends to drop off dramatically once our young people have been Confirmed in the faith. Then we, the adults, complain that "our children are no longer active in the faith" or "we won't see them again until they want to be married." Why do you suppose this is?

I believe that we see many of our young people drift away or disappear, some for a time and others forever, because they have not been invited in and encouraged to connect, to take ownership of the faith in which they have been raised and confirmed. Within the liturgy it is difficult for young people, children especially, to connect because the ritual is adult in its shape and structure. Even so, there is more that we can do for our children, beginning with the smallest, if only to help them feel welcome and to have a sense that they belong. And they do belong, for they are church.

To the few adults who grumbled about our inviting children to gather around the table, I asked them to keep in mind the following:

o   As bread and wine are the symbolic offering of ourselves, let the children be the visible reminder of this.

o   Let their presence be a snapshot of the church as the Body of Christ that is to have life, shape, color, diversity, etc.

o   By helping our children to connect now, we may very well see them remain connected to church. And gone will be the "problem" of our young people disappearing.

I am happy to write that even for some of our grumblers, our children came to be an endearing image of ourselves.

• NO KNEELERS – It was midway through 1998 that the people of St. Joseph adopted the posture of standing during the Eucharistic prayer, remaining in that position until after communion. By the end of the same year the kneelers were removed from the church, not because of our own parishioners, but because of a growing number of visitors to the parish. Unaware of the practice, our guests would instinctively make use of the kneelers. As we know, there is no better way to break the flow of liturgy than the banging of kneelers. And so the decision was made to simply remove the kneelers altogether

For the people of St. Joe's this practice was well received. Sure, we had the usual same few grumblers. This is to be expected. Interestingly, the removal of kneelers was more of an issue for a couple of neighboring parishes than it was for the people of St. Joe's. And while there were some who suggested that "our older parishioners might be upset with this practice," our experience was quite the opposite. It was our "older parishioners" who, in fact, were among the most avid supporters of doing away with the kneelers. To make the point, a woman who was well into her eighties approached me and stated, very directly, "I'm glad someone finally had the b - - - s to do it." Hmmm. Beyond this, a small handful of our older folks posed one question: "What if our health makes it difficult to stand for the Eucharistic prayer?" I responded in this way: "Not a problem. I think God would prefer that you be comfortable. And if that means sitting during the Eucharistic prayer, then so be it." With that, there were no further complaints about our posture, save for those few from other parishes.

- THERESA, AND PREACHING – In recent years some pastors have invited competent leadership persons to preach, on an occasional basis, at the Sunday liturgy. This practice is not at all uncommon within our own diocesan church. This is not about convenience or abdication of responsibility on the part of the ordained. Nor is this a matter of feminism or political equality. It is, however, a practice that is concerned with the business of faith and spiritual growth / development.

  This is what we, as church, have been about for nearly two thousand years. And throughout this time, even to the present moment, have we not heard the voices of countless, competent men and women? In every age all of us, including the ordained, need to hear these voices. And if we, as church, continue to stretch ourselves and listen with ears of faith, then I am confident that together we will hear, through the voice of others, the one voice of the Lord, to which we can only respond AMEN!

With this simple instruction, I then invited Theresa Secord to consider preaching on an occasional basis at St. Joe's. Theresa had grown up in the parish, her own ancestors among the parish's founding members. As "one of their own," Theresa was one of the very first laypersons to be hired by the parish to serve in a capacity other than parish secretary or bookkeeper. For twenty years or more, Theresa Secord wore many hats. And she wore them faithfully as Director of Religious Education, Coordinator of Youth and Adult Ministry and, later, as Pastoral Minister -- all of this within the parish of her youth that had nurtured her faith along the way.

The invitation to preach was gladly accepted. And during her last few years in the parish, before moving on to greener pastures, we were blest to have Theresa preach faithfully and effectively to a people. In and through her voice we must have all heard the one voice of the Lord, for there was only positive feedback from parishioners.

- ALTAR, AMBO and ANOINTING – Within the sacred liturgy there are two primary focal points: The ambo (table of the word) and the altar (table of the Eucharist). The Roman Rite calls for these two objects to compliment one another and to reflect, in construction and design, the dignity of their purpose. Nearly forty years after the Second Vatican Council, many worship spaces fall short in this regard. The parish of St. Joseph was among them.

The old altar and ambo, while they shared an equal space of prominence within the church, were not complimentary of one another. Nor did either one of these objects reflect the dignity of their purpose. The ambo was so small that it was nearly unrecognizable. And both ambo and altar appeared to have been constructed from little more than plywood and second-hand materials. So began the delicate task of educating a people, while at the same time broaching the subject of change, which in the end proved to be both easy and extremely powerful.

With our Liturgy Commission having easily won approval to make a change, the next step was that of actually selecting and purchasing a new ambo and altar. I recall saying to our leadership people that we could simply go the route of buying from a religious goods warehouse. But upon looking through any number of catalogs, we found absolutely nothing that was even remotely appealing to any one of us. Thus we looked for and found an even better route.

A simple postcard drew our attention to a gentleman named Tim Schoonard, who lived on the other side of the state. Advertising as a builder of liturgical objects such as altars and ambos, the parish of St. Joseph called upon Tim. Soon thereafter, he was commissioned to design and build a new altar and ambo for St. Joe's. With a good sense of our worship space and a skill for blending something new with the old, Tim created and delivered a beautiful altar and ambo along with matching candle stands, which included one that could accommodate a Paschal Candle measuring 36 inches tall and 5 inches in diameter.

Our furnishings were delivered during Holy Week of 1998. Now came the matter of anointing and dedicating these objects, a rite that is normally reserved for the Bishop. But it was Holy Week. The Bishop would be busy. And we wanted to do the dedication as part of the Easter Triduum. Returning to the words that have been attributed to St. Augustine, "The church makes the Eucharist and the Eucharist makes the church," we decided that the parish would anoint and dedicate our new ambo and altar. This was done during the Holy Thursday liturgy. And for this special moment we were blest with the presence of Tim Schoonard, along with his wife, Connie, and their seven children.

What made this a key experience for the parish was the very fact that WE did the anointing. I had the privilege of pouring the Sacred Chrism, this richly perfumed oil, upon the ambo and altar. I then invited anyone who so desired to come forth and rub the oil into these objects. Much to my surprise, only a few individuals opted to remain in the pews, and this because they were concerned with having to negotiate a number of steps in order to draw near. What an image! We are church, and we did this sacred action together. An added bonus to this powerful experience was the sweet smell of Chrism that hung in the air for at least a month afterward. We had experienced sacrament in a way that few, if any, of us have ever known.

- MUSIC – In June of 1998 the parish lost its organist. Before having an opportunity to even worry about what we were going to do, there came a knock at the door. Standing before me was a young man named Phillip Konczyk, whom I knew from my home parish. At age 18, having just graduated from Muskegon Catholic Central High School and preparing for college, Phil asked this question: "Are you needing a music director?" My immediate response was to look upward and say aloud "Thank you, God." After a couple of interviews, and having asked the parish to take a risk, Phil was hired not as an organist, but as the Music Director for St. Joe's.

From day one this young man's presence, talent and energy blest the parish. With an incredibly good sense of liturgy and so much more, Phil simply took off. And things were never quite the same again at St. Joe's. The parish was now being stretched musically, and they rose to the occasion. So too was the pastor being stretched, for Phil was quick to remind me that "what is good for the goose...." Together, parish and pastor were continuing to stretch and grow.

As a key experience, the people of St. Joseph came to enjoy the quality, spiritual liveliness and holiness of our praying in song. Whether it was the singing of Eucharistic prayer texts or the pastor and congregation together singing the gathering rite, music had proven to be a most integral part of liturgy. For the people of St. Joe's, and for a growing number of others in Muskegon and beyond, Liturgy was becoming something of that "source, summit and font," of which the Bishops spoke at the Second Vatican Council.

- PERSONAL NOTE – Looking back at four of the most enjoyable years of my life, it is necessary to mention that which I believe was the one thread that held everything together. Without it, I can only wonder if things would have turned out as well. That one thread is this: I was blest to have been in the company of a parish, a people, whose collective mind, heart and spirit were open. This alone opens the door to all kinds of possibilities, the greatest of which is that God can come in. And when God comes in, anything is possible.

- THE EYE OF GOD – At the front of the worship space at St. Joseph, directly behind the altar and ambo, there hung for nearly fifty years a very large work of religious art known as a triptych. This triptych (three panel piece) was a wood carving that had come from Germany and was a depiction of the crucifixion of Christ. It was the belief of many within the parish, including myself, that this particular piece of art, though beautiful in itself, was far too large for a worship space, which, by sharp contrast, was quite small. But that is what we had. And we had to live with it.

  In March of 1999, however, we were alarmed by the discovery that the triptych was pulling away from the wall on which it hung. I say "alarmed" because we were already enjoying the practice of inviting our children to gather around the altar during the liturgy of the Eucharist. Also, directly beneath the triptych were the chairs for both servers and presider. Indeed we were alarmed by the possibility that several hundred pounds of wood might come crashing down at any moment

  Perhaps you have heard stories about individuals who have gone to a yard sale and then returned home to discover that for next to nothing they had just purchased something of a treasure. Like the person who paid only a few dollars for a painting, and then discovered that hiding behind it was a Rembrandt. For the people of St. Joseph, they were about to have a similar experience.

With safety as our first concern, it was determined that the triptych must be removed, at least temporarily. It was early in the week when the workers erected their scaffolding and went about the business of bringing down the triptych. In time there came a knock at the rectory door. I opened the door, and there stood one of the workers who blurted out "Father, you need to come to the church and see what we have just found!"

Entering the church by way of the front sacristy, I first saw the scaffolding, a lot of broken plaster, and dust everywhere. And then I saw it. Light and color were shining through. I turned and looked upward. For a brief moment I was stunned. Then I found the words: "Oh, dear God." We had uncovered a treasure. Looking over us was the eye of God.

Although light had not shown through for nearly fifty years, most of our folks were aware of the existence of the Providence of God stained-glass window which was visible from the outside. With no protective outer covering however, this work of art had fallen victim to the elements and the hand of vandals.

After ooohing and aaahing over their discovery, the workers asked if they should go ahead and cover it over and replaster the entire wall. I then asked them to leave it be, for I wanted parishioners to see for themselves that which had been hidden.

A few days later we celebrated the Fifth Sunday of Lent. At both of our weekend liturgies I was waiting at the rear of the church, long before the first arrival. One-by-one they entered. And one-by-one the response was virtually the same as my own. It seemed that every person, upon entering through the main doors, was stunned. Some stood silent, while others gasped. There was ooohing and aaahing, which turned into a buzz of conversation. And for a good number of our older parishioners, there were tears. These tears were summed up and given expression in the words of an older woman whose entire life had been lived and celebrated within the walls of St. Joe's: "I remember the window. One week it was here, and the next week it was gone. Please, Father, don't tell me we have to cover it up again."

Often I speak to people about God having an incredible sense of timing, and how we need to recapture and appreciate this sense of God's time (Kyros), which is always the right time. This is very different from our typical way of measuring time in terms of days, weeks, months, years, etc. (Chronos, or chronological time). Without a doubt, we were in the midst of one such moment in God's time. This was the right time, the perfect moment for us to have made this great discovery. And for myself, as presider and preacher, it could not have been a more perfect moment.

On this particular weekend we had the joy of hearing the story of the raising of Lazarus from the dead, as told in the eleventh chapter of John's Gospel. The great power and strength of this story is that it serves to instruct us that resurrection is not just something we hope to experience after death, but rather something that is to touch our lives NOW. In other words, to be a Christian is to be a resurrection people, which means that as we make our journey in faith we are to look for, find and celebrate experiences of new life as a way of rehearsing and preparing for the fullness of resurrection in the Kingdom. This was one such occasion for experiencing resurrection in the here and now. The Lazarus story in itself is relatively self-explanatory and on this Fifth Sunday of Lent, even less was required of me as preacher. What I recall most vividly was that immediately following the proclamation of the Gospel, I walked to the front row of pews and shouted: **"REMOVE THE ROCK!"** Then, turning toward the eye of God I shouted: **"COME OUT!"** Already, applause was erupting among the congregation. And once more I shouted: **"UNTIE HIM, AND SET HIM FREE!"** It could not have been any clearer that this was the right time, God's time, for us to have uncovered such a treasure.

By the time we concluded our celebration of Liturgy, and over the course of the next week, there was an overwhelming push on the part of the parish to have this great work of art restored. While a few individuals asked what might become of the old triptych, only one person expressed an interest in having it returned to the worship space. Our next step, then, was to find a person (or persons) who would be capable of repairing, restoring and giving new life to the Providence of God stained-glass window. We would not be disappointed.

One name, Charlie Talbert, came to us from all directions as the one person whom we should approach with our request. In his late seventies, Charlie was well known around Muskegon and beyond. Upon retiring a number of years earlier, Charlie began to pursue the hobby of working with lead and glass. He then discovered and developed his own skill for repairing old and/or damaged stained-glass objects such as lampshades and windows. The rest is history, as Charlie has been called upon by any number of churches and other organizations within Muskegon, and even out of state, all for the purpose of repair and restoration. The call was made, and Charlie responded. Within a very short time, the window was gently removed and then placed into his care.

During the period of restoration, workers returned for the purpose of repairing and resurfacing the interior front wall of the church. We then consulted a liturgist, Fr. Tom Simons, who suggested that we keep it simple, allowing the window to be the primary piece of art on that particular wall. With this in mind, we again called upon the services of Tim Schoonard, the gentleman who had built our new altar and ambo. He, in turn, proceeded to build a processional cross, which fit perfectly with the tables of Word and Eucharist. With all the prep work, etc., complete, we had only to wait.

At the end of August, nearly six months after being placed in the care of Charlie Talbert, the beautifully-restored Providence of God stained-glass window was returned to its place of prominence amid the echo of yet another great story from scripture, that of the prodigal son: What was lost has been found; what was dead has now come back to life.

ABOUT THE PROVIDENCE OF GOD STAINED-GLASS WINDOW

This magnificent window was reclaimed from St. Joseph Catholic Church, Muskegon, just prior to the razing of that structure. Its age is estimated to be more than a century. The diameter is sixty-two inches. Although the window is currently in storage, it is planned to be the only window in the chapel of the Diocesan Priest's Retirement Home, now under construction at 1200 104th Street, SW, Byron Center, MI.

The window is rich in its artistic presentation, vivid colors and Christian symbolism. In examination from the center to the perimeter, one can identify the following:

o   The eye – God the Father
o   The triangle – the Holy Trinity
o   The tongues of fire – the Holy Spirit
o   The circle – the Holy Eucharist, the Son of God
o   The eight squares, each with a cross – the Beatitudes
o   The red encirclement – the ardor of Divine Love
o   The four light blue panels – the Holy Gospels
o   The four contained pearls – the four Evangelists
o   The four outer pearls – the Moral Virtues
o   The perimeter encompassing circle – Eternity

The Providence of God Window will continue to proclaim divine truths revealed by Christ to the Apostles and transmitted faithfully by the Church for all generations.

The Pay-Off

As we began taking risks, trying on change and doing things differently at St. Joe's, the effects were almost immediate. On the down side it came as no surprise to find some persons who were not about to allow themselves to be stretched in their faith. For these few individuals and their families, risk and change were simply not an option. Despite my attempts to reach out, to challenge and yet remain pastorally sensitive, it was ultimately their decision to leave St. Joseph and seek membership within another parish.

By its very nature, growth involves change and letting-go. While change was difficult and even unacceptable for a small number of people at St. Joe's, the overall pay-off was well worth the risk. From 1998-2001 the parish watched and rejoiced as the gaping holes in our congregation disappeared. We were experiencing something of a rebirth and new life. But where were these people, young and old alike, coming from? And why? The response to both of these questions is quite simple.

With regard to increasing numbers, we were able to identify the following sources from which our newcomers were being drawn to St. Joseph:

- The other eight parishes within the Muskegon Deanery. These folks represented something of a lateral movement among local Catholics, which is not uncommon.
- Catholics from outside the Muskegon Deanery. From towns and cities such as Montague, to the north, Ravenna, to the east, Grand Haven and Spring Lake, to the south, people were coming to St. Joseph in downtown Muskegon. This type of movement among Catholics tends to be less common.
- The Order of Christian Initiation of Adults. With the O.C.I.A. in place, the parish was welcoming not just new parishioners, but new members (converts) to the Catholic faith.
- Returning Catholics. This relatively small but significant number of persons represented those who, for whatever reason(s), had found themselves alienated from the church. Inactive in the public celebration of their faith, often for many years, they were invited in and welcomed home at St. Joe's.
- Non-Catholics. While expressing no immediate desire to become Catholic, this small handful of people were likewise being drawn to St. Joseph.

Now for the "why." Why were growing numbers of people finding their way to St. Joe's? On any number of weekends, while greeting people as they were sent forth from Liturgy, I found myself asking such questions as "What's going on here?", or "What is this all about?" Whether they were from another parish, or outside the deanery, or even non-Catholic, their responses to my questions were very telling. And the responses, one after another, echoed a number of common themes:

- "We are fed,"
- "I / We belong,"
- "I / We feel at home at St. Joe's,"
- "I / We have rediscovered/reclaimed our faith."

All of these sentiments came together on a particular weekend when I met a gentleman name Mike, a childhood friend of one of my siblings, and whom I had not seen in many years. After he introduced himself and told me that he was living at Spring Lake, just south of Muskegon, I asked Mike what it was that drew him to St. Joe's for Liturgy. He responded: "The word has gone out that if you want a good experience in Catholic worship, then check out St. Joseph in Muskegon."

So, the word was going out. It seems that we struck a chord, and that we were also proving the wisdom of the dictum "People go where the food is." No doubt, I had a role to play in our experience. And with humility I say that it might possibly have been significant. But mine was no more important than the role that all others played. For it was we, together, who experienced Liturgy as being truly the "work of the people."

At this time I would like to return to music, that element which itself was proven essential to our liturgical celebrations at St. Joseph. From the popular **Gather Comprehensive** hymnal, I wish to share the texts of three songs that had become something of signature pieces at St. Joe's (see appendix). In part, the congregation would sing these particular hymns with resounding energy and spirit, traits that are often lacking in many of our parishes. But the greatest reason for my wanting to share these hymns is that they provide both a reflection and an expression of what it was to BE the people of St. Joseph Catholic Church. And so I invite you, the reader, to prayerfully reflect on the texts <u>ALL ARE WELCOME, WE ARE CALLED and NO LONGER STRANGERS</u>, that you might have a glimpse of who we **were** as a parish, along with a wonderful reminder of what all of us are called to **be** as the larger church.

About Money

Up to this point in our story, little has been said about money. This is very deliberate. I spoke earlier of money as being a necessary part of any enterprise, of a weekly parish income of between twelve hundred and fourteen hundred dollars when I arrived at St. Joe's, and of pastors needing to know that there are appropriate times for writing and/or speaking about financial matters. I mention this because it was our shared belief and our experience that if we are first about the business of feeding people spiritually, money issues will often take care of themselves. Also, not only will the money come, the parish might actually find itself enjoying a growing surplus. This too was our experience. And so a reminder, especially to pastors: Don't begin and end with the buck. Instead, begin and end with matters of faith and spirituality. The rest will take care of itself.

Chapter Two
AN UNEXPECTED TURN

The Problem

On the weekend of October 14/15 of the year 2000, I had the sad task of informing the parish of a potential problem of large-scale proportion. It was discovered, and shortly thereafter confirmed, that the south wall of our old church was moving outward, literally pulling away from the rest of the structure. The cause of this shifting, we would later learn, was the result of the failing support structures that held up the roof of our 117 year-old church building. With age and stress having caught up with the structure, it was apparent that time and circumstance had yet one more role to play in our story.

A Different and Difficult Process

Having alerted the parish to a problem of major concern, the next step was to ask the faith community "What are we to do?" and "Where do we go with all of this?" These questions would be presented the following weekend, October 21/22. But first, there were a couple of matters of immediate concern:

- The Paschal Candle (Easter Candle) was placed prominently at the front of our worship space. There it would remain as the visible reminder that the Risen Christ, that Light which no darkness can overcome, is always with us on our journey.
- Words were spoken, and a written statement presented, assuring the parish that our old church was not in danger of collapse. Knowing that we could remain "at home", at least for a time, would prove to be helpful and most comforting for all of us at St. Joe's.

In response to the questions "What are we to do?" and "Where do we go with all of this?", I shared with the parish my belief that any decisions to be made should not fall either to myself or to the Bishop. Instead, I found myself saying to the people of St. Joe's, "**We** must wrestle with a number of issues, and together **we** must come to a decision about ourselves." Of course, we would keep the Bishop informed every step of the way. And with that said, I then invited the parish to roll up their pant-legs and get ready to wade into some muddy waters.

With parish leadership bodies gathered as one, they were given the task of leading the people of St. Joseph through a difficult and demanding process. There was much more to look at than the immediate problem of the south wall of a building. All of our facilities were aged and in need of attention. And beyond our own needs, the parish also found itself in the midst of a deanery-wide planning process in which all nine parishes were looking at how to meet the future needs of Muskegon's Catholic community with a rapidly-shrinking number of priests. With all of this before us, the parish entered into the waters of uncertainty. But we did so with much attention given to the following:

- First and foremost, the need to remain firmly rooted in faith as we discerned **where** and to **what** the Lord might be calling us in this unique situation.
- Returning to the operating principles that were developed earlier in our journey (see pages 8-11).
- The need for pastor and leadership persons to be up front, honest, and direct with parishioners every step of the way in the discernment process.
- Inviting, encouraging and even pushing the parish, if necessary, to take ownership of the process, stressing that we as a community of believers would soon determine the very fate of our parish. To best facilitate parish ownership of this process, pastor and leadership persons would have to keep parishioners informed about such things as: structural analysis of the church building and recommended action, the needs assessment of all buildings and grounds (looking at the most immediate needs, such as new windows, wiring and plumbing upgrades to the buildings) and of course, the projected costs.
- Keeping before us the reminder that as church we are larger than any one of us, and larger than any parish.

- The need to ask ourselves, repeatedly, throughout the process: Is it good and faithful stewardship to continue putting large sums of money into old buildings simply because of our sentimental attachment to them?
- And finally, the invitation for all parishioners to prayerfully reflect on the words of our parish mission statement:

WE THE PEOPLE OF THE ROMAN CATHOLIC PARISH OF ST. JOSEPH SEE OURSELVES AS THE PEOPLE OF GOD. WE DEDICATE OURSELVES TO PROCLAIMING AND LIVING THE GOSPEL OF JESUS CHRIST. IN THAT PROCESS WE STRIVE TO BUILD A COMMUNITY OF FAITH WHERE EACH PERSON IS VALUED AS A CHILD OF GOD. EMPOWERED BY THE HOLY SPIRIT AND STRENGTHENED BY OUR PARTICIPATION IN PRAYER AND THE SACRAMENTS, WE ENDEAVOR TO CONTINUE THE SAVING WORK OF JESUS IN OUR WORLD.

For nearly four months, things remained quite normal at St. Joe's. We were church, and so our attitude and approach were nothing less than "business as usual." But during this time we were also praying over and wrestling with some very large issues. And as the dollar figures started coming in, a rather daunting picture began to emerge. This is reflected in the accompanying insert, ST. JOSEPH FACILITIES NEEDS.

ST. JOSEPH FACILITIES

Listed below are any number of areas in which our aging facilities are – or will soon be – in need. This is not an all-inclusive list, but with at least one bid having come in on each of the higher dollar items, we are now given a glimpse of the larger picture that is before us.

ITEM	BID #1		BID #2	
Electrical	* $ 3,800.00	(church)		
	* $ 9,250.00	(rectory)		
Exterior brickwork	* $ 49,879.00	(church)		
Interior painting	* $ 7,000.00	(rectory)	$14,185.00	combined bid
	* $ 10,600.00	(church)		
Carpet / flooring	* $ 7,199.00	(rectory)		
	* $ 11,915.00	(church)		
** Windows	* $ 13,575.00	(rectory)	$12,300.00	
Face restoration (exterior wood)	* $ 25,560.00	(church)		
Parking lot (behind family center)	* $ 10,638.00		$15,800.00	
Main parking lot	$ 2,690.00 - * $3,000.00			
Structural analysis of the church	* $ 7,200.00		$3,000.00	
Church wall / roof	* $125,000.00 - $150,000.00			

$284,616.00 (combined sum of those figures which have a * to the left of the dollar sign)

** For a ballpark figure that would include new windows in the family center, both companies suggested we simply double their respective bids.

$284,616.00
+ $ 13,575.00   (windows in family   center)
+ $ 32,274.00   (INL CAMPAIGN)
+ $ 2,500.00   (Conventional Baptismal
                Font)
TOTAL $332,965.00

The Decision

On February 11, 2001, the people of St. Joe's gathered for Sunday liturgy. But this was no ordinary celebration, for this was the day of decision. Up to this point, people had been informed of all that was going on during the previous four months: The structural analysis of our old church, the various other needs of our aging facilities, potential costs, etc., all of which had drawn us to look very critically at ourselves, the Muskegon Pastoral Planning Process and our relationship to the larger Catholic church. This was the framework within which the parish would have to make the most difficult decision of their 117-year history.

Immediately following our liturgy, the people of St. Joe's remained in the church. Various members of our leadership bodies then proceeded with a brief review of the process, of all the information and the projected costs for keeping the parish open. Along the way a local business consultant further advised us, that if our decision were to move forward with making a large investment in our parish and its aging buildings, we should also include raising an additional minimum $50,000.00 as a growth fund for the ongoing care and maintenance of the facilities. At this point our building needs were fast approaching a projected cost of nearly $500,000.00.

With the review behind us, I stood before the people of St. Joseph and asked this question: "Is there anyone among us who believes that we should pull all the stops and proceed with a major capital campaign?" Not a single hand was raised. Not one word was spoken. For a moment there was only silence – and then the tears began to fall.

Amid the tears, the next painful step was for me to present the parish leaderships' recommendation that we approach our Bishop with the request that he close the parish of St. Joseph. Barely able to get these words out of my mouth, something most precious then occurred. A few of our senior members, whose entire lives had been connected to and celebrated within this parish, came forward and asked if they could share some words. Each of them said much the same. But the most eloquent words came from Joe. In his late seventies and crippled since birth, Joe limped to the front of the church and spoke these words: "I never thought that I would even think these words, much less say them, but it's time. It's time to let go."

"Time to let go." These words led us to seek and achieve an overwhelming consensus to accept the parish leaderships' recommendation. With that, each of our leadership persons then signed their names to the letter of recommendation that was then presented to Bishop Robert J. Rose. But the people of St. Joseph went an extra step in this procedure. One by one, a large number of our parishioners came forward to place their names on a simple statement, which indicated that they were in agreement with the recommendation that our parish be closed.

Letting Go

With the decision made, Bishop Rose accepted the parish's recommendation of closure. Because the diocese had never closed a parish, there was no blueprint for us to follow. And so the Bishop allowed us great latitude in mapping out the final steps in the life of St. Joe's.

Our first order of business was to target the date on which we would gather for the closing liturgy. Initially, it was our desire to look at the date of March 19th, the Feast of St. Joseph, as the most appropriate day for bidding farewell to the parish. However, the Bishop could not be with us on that particular date, and so we moved our closing liturgy to March 25th because we wanted and needed our Bishop to lead us in prayer on this most solemn occasion.

With the date set for the closing liturgy, it was also determined that the parish begin the process of letting go and separating out. To facilitate this process, Muskegon's Catholic Central High School graciously opened their doors to the people of St. Joseph. There we would gather for weekly liturgy, continuing our journey together until we assembled one last time within the walls of St. Joe's on March 25th.

As we began planning for the closing of the parish, for an end of sorts, we were given another experience of Kyros time, of "God's time," which is always the "right time." This difficult part of our journey would unfold during the season of Lent, which proved spiritually beneficial for all of us. What follows are some bulletin excerpts from the weekends of February 26, March 4, and March 11, as penned by the pastor.

Harpe Melodies

2/25/01

Come Wednesday we will again mark ourselves with ashes and enter into our annual observance of Lent....our time of emptying and letting go so that we may come to Easter, or could it be so that Easter may come to us. One of the most familiar Lenten images is that of the grain of wheat: "Unless the grain of wheat falls to the earth and dies, it remains just a grain of wheat. But if it dies, it produces much fruit." (John 12:24) I would like to invite and encourage every member of St. Joe's to pray this scripture each and every day of Lent, for we are that grain. Fr. Dave

3/4/01

This past Tuesday the liturgy presented us with a reading from the book of Sirach (35:1-12) that prepares us well for welcoming the season of Lent. The thrust of the entire reading is captured in verse 9: "Give to the Most High as He has given to you, generously, according to your means." Too often we think of Lent in terms of giving up something, only to take it back at Easter. And for some, their "giving" or "giving up" is accompanied by bitterness. Sirach reminds us that such grudging and forced giving is not giving at all. The words of Sirach hold true for the Christian: Our Lenten giving is to be accompanied by generosity, freedom, cheerful countenance and a spirit of joy. So, why is it that some still approach Lent with a spirit of negativity, as if it were a downtime of drudgery? Perhaps there is some ungrounded fear that any true giving or giving up will somehow diminish us. On the contrary, our giving during Lent is what leads us to Easter. And our giving during this season (and always) simply makes us more Christ-like.

For the people of St. Joseph and for the larger Catholic community of Muskegon, this year's Lenten journey holds great promise. As we move toward closing some doors and opening others, I invite all to be aware of the desert themes of Lent: Hungry and helpless, yet called to remain faithful, God's promise of a glorious future, Gathering at the water, Seeing more clearly, Having to die in order to rise, Facing the buffets and spitting. And when we arrive at Easter, be aware of its various themes: New beginning after hope was gone, The church earning the respect of the community, Standing up to the world, Growing and spreading out in the midst of a hostile world, Ministry and outreach, United (always) around what is central.

Just a few words of a business nature....On Sunday, March 25, at 6:00 pm we will gather in the church. There, in the company of Bishop Robert Rose, we will celebrate our closing Liturgy. Between now and March 25, I hope to do a general mailing to all registered parishioners. In that letter I will address a number of concerns relative to the buildings, contents, etc., and what is to become of them. So stay tuned.

In the meantime, let us all be of good cheer, generosity and joy during this season of Lent....and always. Fr. Dave

3/11/01

Once again, the Lenten scriptures offer us nothing less than hope, promise and a new beginning. As we bring a parish to closure and move beyond the walls of St. Joseph, the readings for this Second Sunday of Lent present us with God's promise of a glorious future. We may be in a moment of sadness, but we have all that we need to sustain us as we go forth, namely, God's Word and God's promise. Therefore, we have, as the psalmist reminds us, nothing to fear.

May the Lord BE our light and salvation....always. Fr. Dave

It was previously stated that the Diocese of Grand Rapids had never before closed a parish. As such, our experience would be a first for the people of the parish, for myself, a priest for less than seven years, and for the entire diocesan church. Fast approaching the date for our closing Liturgy, a number of questions emerged with regard to practical concerns such as sacramental records, building contents, etc., and what will become of these after March 25th. The most effective way of responding to such questions, all of them valid, was by way of a final pastoral letter to the parish. Dated March 12, 2001, that letter appears here in its entirety, although there were some changes that will be addressed shortly, under the heading Giving It Away.

Accompanying the general mailing to each of our parishioners was a simple prayer card. On one side of the card there was a color print, a depiction of Jesus washing Peter's feet. On the other side was the Prayer of St. Francis of Assisi:

> Lord, make me an instrument of your peace.
> Where there is hatred, let me sow love.
> Where there is injury, pardon.
> Where there is doubt, faith.
> Where there is despair, hope.
> Where there is darkness, light.
> And where there is sadness, joy.
>
> O Divine Master, grant that I may not so much seek
> to be consoled as to console,
> to be understood as to understand,
> to be loved as to love.
> For it is in giving that we receive.
> It is in pardoning that we are pardoned.
> And it is in dying that we are born to eternal life.

ST. JOSEPH PARISH
1224 Fifth Street
Muskegon, MI 49441
(231) 722-2556

12 March 2001

Dear friend and parishioner of St. Joseph,

Greetings in Christ. We are witnessing and participating in a historic moment that reaches beyond ourselves, touching many people, Catholic and non-Catholic alike. In recent weeks, as we have been preparing for closure, a number of persons have called to inquire about a host of parish-related concerns which go beyond March 25th. And so I thought it best to effectively address these concerns by way of a general mailing to all the registered parishioners, something of a final pastoral letter, if you will.

First, here are a couple of dates to remember:

SATURDAY, MARCH 24th, at 5:30 pm – Potluck Dinner in the family center. Gloria Lack has graciously volunteered to coordinate the dinner. If you have questions, or if you wish to be of assistance, please call Gloria.

SUNDAY, MARCH 25th, at 6:00 pm – Closing Liturgy at St. Joseph Catholic Church. As we are anticipating a capacity crowd, you are advised to plan on arriving well in advance of 6:00 pm. Also, some have expressed concern about not wanting to drive into town for an evening Liturgy. Not to worry. Patty Siewert has volunteered to coordinate transportation. So, if you are in need of a ride on the 25th, please call Patty.

Now, things to come after March 25th:

- SACRAMENTAL RECORDS will be in the care of Our Lady of Grace Parish, Muskegon.
- EYE OF GOD STAINED-GLASS WINDOW will be removed, given a protective covering and then be placed in

the custodial care of Our Lady of Grace Parish, Muskegon. It is hoped that this work of art might some day be incorporated in a new structure. That would be wonderful.

- ALTAR, AMBO, CANDLESTANDS, PROCESSIONAL CROSS, STATIONS OF THE CROSS, PRESIDER AND SERVER'S CHAIRS will be making their way over to Our Lady of Grace. For our parishioners who are connecting with Our Lady of Grace, you bring these incredible gifts from one parish to another. What a visible reminder you will have of St. Joe's every time you walk through the doors at Our Lady of Grace.

- HYMNALS – Like God's Word, the Bible, hymnals need to be opened up, not stored up. I've had one person ask if they could take the hymnals which they purchased and donate them to another parish. That's fine with me. Otherwise, Our Lady of Grace is entertaining thoughts of purchasing the **Gather** hymnal, and they have expressed interest in acquiring ours. Again, a great gift and a wonderful way of honoring the memory of this parish.

- STATEMENTS OF GIVING – Our administrative assistant, Joe Secord, has said that in a timely fashion he will mail out statements of giving for the year 2001 following the closing of the parish.

- CLOSING OF THE FINANCIAL BOOKS – We have targeted June 30th, the last day of the fiscal year, as the point at which the parish's financial books will be permanently closed. This may or may not be impacted by #9 below (buildings and grounds).

- CATHOLIC SERVICES APPEAL – There will be no CSA for St. Joseph Parish, as a result of our closing. As part of the larger diocesan church, however, I invite and encourage you to continue participating in CSA, pledging to whatever parish you will be attending.

- CATHOLIC SCHOOLS – The parish will honor its assessment to Greater Muskegon Catholic Schools for the current school year.

- BUILDINGS AND GROUNDS – During the latter part of April we will hold an estate-type sale, open first to the people of the parish, then to the general public. As for the buildings? Questions of loss prevention and liability have led us to expedite the process of having the buildings razed, with the church being the first to come down. While the buildings are insured through June 30th, it behooves us to act in a responsible fashion. For as we know, buildings that are unused – whether they are insured or not – often invite unwanted activity and/or visitors. Again, prudence as well as the question of liability must dictate our action.
- IN THE NAME OF THE LORD CAMPAIGN – We still have a balance (outstanding) of nearly $33,000. to the diocese's INL campaign. As we close the parish, it would be wonderful to know that we have honored all of our obligations. As we are finding out, it costs to close a parish. And so we have deliberately placed INL at the bottom of the list of priorities as we move through this process.
- REMAINING ASSETS – The question has been raised "What if there are any remaining financial assets after all is said and done?" Mr. John Czachorski, diocesan finance director, and Bishop Rose have offered a couple of options for our consideration in this regard:
  - A donation, in the name of the parish, to some local cause/charity.

  Or…

  - If there is a substantial sum of money remaining, the diocese would hold it in trust. Then, if down the road we were to see a building rise from the ashes on this holy ground, the monies would then be returned. Such a building would not be a parish, but rather a multi-use facility whereby we , and a host of others, could maintain a Christian presence in this part of downtown Muskegon. Hmmm. Just imagine the possibilities

- BOOK: <u>WHERE THE STAR CAME TO REST: HISTORY OF THE DIOCESE OF GRAND RAPIDS, by Monsignor Gaspar Ancona</u> - I recently spoke about this book, referred to as the Magi Project. Parishes are already in (or will soon be in) the process of taking orders for this book, the first part of which is the diocesan history, followed by a picture and brief history of each parish within our diocese. We will not be taking orders through St. Joseph Parish. If you are interested in purchasing the book, cost of which is $24.95, you may exercise one of these options:

  Order through the parish you will be attending,
  Or through the diocesan bookstore, by mail or phone:

  > MAGI PROJECT
  > Bishop Baraga Bookstore
  > 600 Burton Street SE
  > Grand Rapids, MI 49507
  > Phone: (616) 245-2251

As it is time to bring this letter to a close, I think it is appropriate to do so on a personal note. In the nearly four years that I have been at St. Joe's, I have witnessed the incredible faith of a people. It is the faith that birthed this parish, the faith that sustained it for over 117 years, and the same faith that will endure beyond its closing. In all of this I recall the occasion of my priesthood ordination, at which time a newspaper reporter asked: "What do you envision priesthood to be?" My response, now, is the same as it was then: "I envision priesthood as part of a people's faith journey. As such, I want only to be a good priest, walking with people and stopping along the way to celebrate the sacred moments in their lives." Thank you for this sacred moment, one that has been filled with many.

May the Lord bless each member of this family, St. Joseph, with faith, hope, love, good health, countless friends, much laughter, music, and a spirit that knows no bounds.

Peace.... Always,
Fr. Dave

Potluck Meal

> On the first day of unleavened bread, when they had killed the Passover, his disciples said to him "Where shall we go to prepare the Passover?" And he sent forth two of his disciples, instructing them: "Go into the city, and there you will meet a man carrying a water jar. Follow him. And whatever house he enters, say to him 'The Master says where is the guest room, where I shall eat the Passover with my disciples?' And he will show you a large upper room, furnished and prepared."
>
> Mark 14:12-15

On the eve of his passion and death Jesus gathered with his closest friends to share one last meal. On the eve of our passion and death as a parish, inspired by Jesus and the words of the Evangelist, the people of St. Joseph gathered to share one last meal. The large upper room of our parish family center was furnished and prepared. There we gathered to feast, to remember and celebrate, all of which was accompanied by a wonderful mix of laughter and tears. By the time we dispersed, it was late into the evening. And darkness covered everything.

Closing Liturgy

The first Liturgy at St. Joseph was celebrated on Christmas Day, 1883. The odds are that when the doors were first opened there was snow on the ground. How appropriate that 117 years and 3 months later, to the day, there would be snow on the ground as a people gathered one last time for the purpose of closing the same doors.

At 6:00 pm Sunday, March 25th, 2001, the people of St. Joe's celebrated their final Liturgy. We were not alone in this experience. Newspaper writers and local television crews were present to witness this event that was a loss not only for the Catholic community, but the larger community of Muskegon as well as the eleven-county diocese of Grand Rapids. Also invited were the pastors of Muskegon's eight other Catholic parishes. And of course, leading us in prayer on this most solemn occasion was our Bishop, Robert J. Rose.

No amount of words can capture this incredible and faith-filled celebration. And it was, in every sense of the word, a celebration. But we have been left with the next best thing, something of that experience for each of us to hold on to. One of our parishioners, David Loring, along with his wife, Lori, created a video, which was later mass-produced. This video consists of two parts: A well-narrated history of the parish with photos and video clips of times past, followed by the closing Liturgy in its entirety. Within a matter of months after closing the parish, I was told that hundreds of these videos had gone forth from Dave's studio and that he was continuing to take orders for still more. And based on the words of those who purchased the video, it seems that our story has already traveled north, south, east and west around the U.S.

While impossible to articulate the experience of our closing Liturgy, it is very possible to highlight those elements that added considerable shape and substance to such a beautiful moment:

- A worship aid was printed for this special occasion. Containing all the music that would be sung, as well as an outline for the Liturgy, the program cover was adorned with a color print of the Providence of God stained-glass window that was reclaimed and restored in 1999.
- There was a prevailing sense, in our planning, that the closing Liturgy should be much like our normal Sunday gatherings. And so it was. With the exception of the presence of the Bishop, neighboring priests and invited guests of the media, this Liturgy was much the same as so many others that we had celebrated in recent years. There were those key elements of liveliness of music and spirit, incense, bread that was made by our own hands, and of course our children gathered around the altar.
- On this occasion, as on so many others, the music was superb. Very deliberately, we chose as our gathering song the Litany of the Saints, asking the holy men and women of all time to pray for the people of St. Joe's during this difficult moment and beyond. And near the end of our celebration, it was in music that we were given yet another treasure as our

music ministers presented us with a song, <u>BEYOND THIS DAY</u>, by our own Phil Konczyk. Written specifically for our closing Liturgy, the text appears in the appendix to this book.

- This Liturgy was truly a celebration of praise and gratitude to God for all that had transpired within the walls of this sacred space, St. Joseph: 117 years of a people's call to holiness, a journey that was marked by a seemingly countless number of sacred moments such as Baptisms, First Eucharist celebrations, Reconciliation to God and to one another, Confirmations, Weddings and Anointings. 117 years of welcoming new life, celebrating and nurturing that life, and ultimately letting go and giving it back to God.

  Beyond what has just been written, I would like to share a few personal vignettes surrounding the closing Liturgy at St. Joe's

- Throughout the Liturgy it seemed apparent that the Bishop was visibly moved by our celebration of prayer and praise. After its conclusion, and having greeted many tearful parishioners as they departed, the Bishop embraced me and said: "This was superb."

- Sr. Patrice Konwinski, Chancellor of the Diocese of Grand Rapids, was also present at our closing Liturgy. As our primary contact person (at the diocesan level) throughout the decision-making process, it was only fitting that she would attend this particular event. Her presence, for many of us, was warming. And as she left the building, Patrice said to me: "David, a stranger could have walked in and never guessed that this was a sad occasion, not judging by the life and energy within these walls tonight."

- Among the very last to leave the building was a young family, Jim and Patty Siewert, along with their son, Kurtis. They waited patiently, as I turned down the lights and prepared to lock the church, for they wished to present me with a gift. What I unwrapped was no ordinary gift, but a real treasure. Matted, framed and ready for hanging was a beautiful color painting of the exterior of St. Joseph Catholic Church. This gem, this work of art, had come not only from the heart of the Siewert family, but from the very heart and hand of Patty's own father who, incidentally, was not even one of my parishioners.

- Finally, upon returning to the rectory, I listened to the lone message on the answering machine. It was the voice of my brother, Gary. Calling from his home in Detroit, at the precise time that our closing Liturgy was to begin, his message was as follows: "This is Gary and Anne. We, along with the kids (Emily and Ben), want you and the parish to know that we are praying for all of you, as St. Joseph performs its swan song. We love you."

Late into the night, as I drifted off to sleep with all these things in my heart, I began to sense that it was not yet complete. Perhaps the "swan song" had just begun.

Giving It Away

As he was setting out on a journey a man came running up, knelt down before him and asked, "Good Teacher, what must I do to share in everlasting life?" Jesus answered, "Why do you call me good? No one is good but God alone. You know the commandments:

'You shall not kill;
You shall not commit adultery;
You shall not steal;
You shall not bear false witness;
You shall not defraud;
Honor your father and your mother.'"

He replied, "Teacher, I have kept all these since my childhood." Then Jesus looked at him with love and told him, "There is one thing more you must do. Go and sell what you have and give to the poor; you will then have treasure in heaven. After that, come and follow me." At these words the man's face fell. He went away sad, for he had many possessions.

Mark 10:17-22

With the parish now officially closed, the people of St. Joseph would have to go about the business of checking out the surrounding parishes in search of a new home. This would prove to be relatively easy for some, but difficult for many including myself. No longer a pastor, at least for the next few months, I was now an administrator whose task it was to oversee the emptying of buildings and their ultimate demise.

With regard to the emptying of buildings, the initial plan was to have an estate sale of sorts, as mentioned in the pastoral letter of March 12th, 2001. However, a number of persons expressed concern about the need for a sale. "Why sell everything?", they asked; "we are no longer a parish." And so it came to be that after polling a number of persons and having found inspiration in the words of the Evangelist, Mark, the decision was made to simply give it all away. In the process there were a few persons who felt obliged to give something in return for the objects they wished to acquire from St. Joe's. To these individuals it was suggested they simply consider a monetary contribution to a local charity in the name and memory of St. Joseph Parish.

While it is impossible to capture the fullness of the experience of "giving it away," I have found the following to be particularly noteworthy:

RECTORY AND FAMILY CENTER

Clearing out the parish rectory and family center was a relatively easy task. For myself as well as a few others, it was as if we were called upon to play the role of Santa Claus, simply giving things away right and left. Our only wish or desire was that all furniture and household items go to individuals and/or families who found themselves in need. To my knowledge, that wish was fulfilled. But our giving didn't stop there.

A number of persons, including some outsiders, expressed interest in much of the old woodwork, especially door trimmings, baseboards and the multitude of five-panel doors within the buildings. Among these individuals was a young married couple,

Jon and Carey. Just one block away from St. Joe's, they were in the process of restoring (to period) the old house that was now their home. Having observed them carefully documenting each and every piece they were removing from our buildings, I was prompted to ask them "why?" They responded: "For ourselves and for anyone who might live in our house after we are gone, we want a visible reminder that these things came from St. Joseph Catholic Church."

CHURCH

As we readied the buildings for demolition, it seemed that our old church had some parting gifts for the people who had gathered within its walls for over 117 years. There were even a few gifts that went out beyond the people of St. Joe's.

To parishioners, the church pews were available, as were portions of the colored glass windows from the sidewalls of the church. Even the redwood ceiling panels were carefully removed. This was done by a local contractor who had assisted his own father with the installation of that very ceiling some thirty years before.

In the end, memories alone would be sufficient for many parishioners. For others, however, a token of sorts would be helpful, perhaps even necessary, as a tangible reminder of what had been.

Reaching beyond our own walls, the following gifts were sent forth from St. Joe's:

- Our hymnals, purchased by parishioners and embossed with the name of our parish, were given to a Catholic church at White Cloud, Michigan that is also part of the diocese of Grand Rapids. Many of us have since wondered if it was more than coincidental that this particular gift would find its way into the hands of another parish named St. Joseph.

- The church's cornerstone, along with the bell that had called so many to prayer and had tolled for seemingly countless deaths, were given to the larger community of Muskegon. These gifts are now in the care of the local museum.

- To the entire diocesan church, which is comprised of eleven West Michigan counties, the people of St. Joe's left two gifts of particular significance. First, the contents of a time capsule that had been placed with the laying of the church cornerstone. Much like the very people themselves, these items were simple: a handful of coins, along with a few newspaper articles from the year 1883.

The second gift was a four-piece set of liturgical vessels that included a chalice, ciborium, paten and monstrance. Dated to the late nineteenth or early twentieth century, and nearly oozing with gold plating as well as diamonds and rubies, these items reflected something of a bygone era within Catholicism. The vessels were presented to Fr. Nicholas Irmen, by one of his friends, as a gift to the priest and his parish, St. Joseph, Muskegon, which Irmen served from 1891-1911. For obvious reasons, these items would be used only on special occasions such as Christmas and Easter.

In 1986 the pastor of St. Joe's, then Fr. Bernard Hall, questioned the wisdom of keeping these particular vessels on the grounds of the parish. An appraisal of the four items placed their combined worth at nearly five hundred thousand dollars. With concern for their safe keeping as well as the exorbitant cost of insuring these items against loss or theft, the pastor and parish leadership persons entered into a most amicable arrangement with the Muskegon Museum. The vessels were placed on indefinite loan to the museum which, in turn, insured the items, while at the same time reserving the right to routinely display them along side other pieces or works of religious art. The "sacred vessels," as they were fondly referred to, remained in the care of the local museum until September of the year 2000. Shortly thereafter, driven by the same concerns as those of Fr. Bernard, the items were transferred to Grand Rapids, where they are now tucked away in the diocesan archives.

- Finally, there is one gift that has yet to be given. With all of the parish's financial obligations met, there was nearly twenty thousand dollars remaining in our savings account in Grand Rapids. Already, there was some small talk going on in Muskegon, at an ecumenical level, about the possibility of redeveloping the site of St. Joe's and thus maintaining a religious presence in the neighborhood. And so this proposal was accepted by Bishop Rose: The remaining financial assets of St. Joe's are to go toward that site redevelopment, if such a plan is materialized within the near future. If not, then there is to be one donation to a local Muskegon charity in the name and memory of St. Joseph Catholic Church.

A Bright Moment

With the parish closed and the buildings emptied and awaiting demolition, the next order of business was for the people of St. Joe's to begin looking at the surrounding parishes, to find a new home within the larger faith community. For some, this was a relatively easy task. For many of us, however, the experience can only be described as painfully difficult. This was not a typical moment of transition. Ours was a death experience. Having just lost something special, much like the passing of a family member or beloved friend, one cannot begin to "move on" until one has first taken the time to mourn their loss. Only then does the healing process begin. And only then is one able to effectively adjust to that which is gone forever.

Adding to our pain and difficulty was the growing perception that diocesan leadership persons were very insensitive to the needs of a people who had suffered a great loss and who now found themselves hurting. In the hearts and minds of many, including myself, it began to seem as if the parish's journey of the past seven or eight months was treated at the diocesan level as simply a matter of business, with little concern for the human lives that were involved. True to form, however, the people of St. Joe's would rise above even that experience.

Aware of both our pain and the need to remain with one another, at least for a time, something of life and beauty unfolded before us. With the new life of spring arriving in West Michigan, Gary and Mary Jeanne Silvis asked if I would consider gathering with the people of St. Joe's and celebrating Sunday Liturgy in the backyard of their home. Of course, I said "yes" without hesitation. However

I made the request that we invite Phil Konczyk to grace us with his presence as well as his music skills.

Over the course of the next few months a people gathered, regularly, for outdoor Liturgy. The last Paschal Candle from St. Joseph was present, continuing to provide light in a moment of pain and uncertainty. Red wine and loaves of unleavened bread were presented in abundance. There was laughter and tears, music and song. We remained, for we were church **before and beyond** the walls of St. Joseph, or any other church building for that matter. But something was noticeably different. It was not just the people of St. Joe's who were gathering at "St. Silvis," as we affectionately referred to ourselves. Many unfamiliar faces were present as well. All were welcome. And together, in the company of our God, all would find some much-needed comfort, strength, hope and nourishment for the continuing journey of faith.

As our time together drew to a close, and as I prepared for a new assignment within the diocese, two of our final gatherings would become forever etched in the hearts of many. First, there was our celebration of Pentecost Sunday, 2001. It was a gray and chilly morning in Muskegon. With rain in the forecast, I began to wonder if we would have to forego this particular Liturgy at St. Silvis. But the people of God arrived in relatively large numbers, ready to celebrate. As we gathered, the clouds began to scatter and the sun shined upon us, accompanied by a soft breeze. We needed this time together, and the Lord knew it. The rain would return, but not until our celebration was complete. With no walls and doors keeping us in, or keeping others out, this would become for many of us a most treasured Pentecost.

Given all that the parish was grappling with, on this particular Pentecost Sunday I returned to a prayer that I had shared with the people of St. Joe's the previous year. On behalf of the parish of St. Joseph, I would like to share the prayer of the late Archbishop Thomas J. Murphy, just as it was presented to them on June 25th, 2000:

Harpe Melodies

6/25/00

In April of 1997 Archbishop Thomas J. Murphy of Seattle was dying of leukemia. As part of his farewell to his people he allowed a local TV station to interview him about his life and beliefs. The final question asked of Archbishop Murphy was simply, "Are you afraid to die?"

In his prophetic way he proceeded to tell a story about a young person who impacted his life and ministry.

"When I was bishop in Montana I had a special meeting with high school aged people who were preparing for their Confirmation. I asked what I thought was a basic yet profound question of the group. 'What would you die for?'"

In the back of the room a hand went up and a 15-year old young man respectfully rose and said, "Bishop, I think you're asking the wrong question. The question isn't what I would die for – it's what would I **live** for."

Archbishop Murphy carried that question with him through his final illness and into eternity. He lived for Christ.

We've just celebrated the great feast of Pentecost. As a people who are now called to bring the Easter experience to the events of ordinary time and life, I invite all to reflect on the words of the late Archbishop. Peace always. Fr. Dave.

<u>A NEW WIND AND FIRE</u>
"Prayer for the New Millennium"
for the Faithful in Seattle and throughout the country
by Archbishop Thomas J. Murphy

We are a blessed people.
We live on the edge of a continent
in a vibrant country of beauty and grandeur
reflecting the goodness of a gracious God.

We are a people of faith, and yet, we are
dreamers and idealists,
realists and pragmatists,
people of doubt and people of hope.

We live in the shadow of the end of one century,
and we are poised on the edge of a new millennium.
We are aware of the sin and grace
that are part of the human condition.

We are not unlike the men and women
who gathered in the upper room centuries ago
to listen, to wait, to pray.

They were afraid, uncertain of the future.
But they remembered the promise that Jesus made.
Jesus promised a spirit of love, of truth, of hope
that would change their lives.

It was early morning when the wind and fire came.
The Holy Spirit embraced the earth and the people,
and the followers of Jesus would never be the same.

As we approach a new millennium,
a time of grace, a time of hope,
a time of passion and promise....
It is time again to embrace the wind and fire
of the first Pentecost.

The second experience that came to be etched within our hearts
was our final gathering for an outdoor Liturgy and potluck meal at
Muskegon's McGraft Park. This was to be, primarily, a time for
saying good-bye to one another. But it was also an occasion for still
more giving. We had told Phil Konczyk that a rented piano would
be at our disposal for this particular Liturgy. At the conclusion of
our celebration, I spoke these words directly to Phil: "I must confess
that I have lied to you, Phil. This brand new piano is not a rental.

Identical to the very one that you had selected for the parish, this one now belongs to you. It is a gift from the people of St. Joe's. And with the use of the rental truck that is here, we will deliver the piano to your home this evening." With Phil already in tears, the people of St. Joe's then came forward to bless both Phil and his new piano with their own tears and words of love and encouragement.

Just moments after presenting Phil with a new piano, it would be my unexpected good fortune to become the recipient. Darrell and Marcia Senior, along with their children, Nick and Rachel, came forward. Catching me somewhat off guard, they placed a gift into my hands and said: "With love, Fr. Dave, and with the assistance of a local artist, the people of St. Joe's wish to present you with this gift." Opening the box, I discovered a circular sun-catcher that was composed of rich colored glass and held together by lead framing. And as I removed this piece of art, exposing it to the light, I was immediately taken back to a very special day in the month of March, 1999: Light and color were shining through. And looking over us was the Eye of God. Although this particular piece was not a replica of The Providence of God stained-glass window, it would serve as a beautiful reminder of a powerful experience that had been shared by many.

At the end of the evening, as we prepared to go our separate ways, there was yet another gift for all who were present. It was the following prayer, most appropriate not only for our experience as a parish, but also for all who journey the path of faith.

PILGRIM PRAYER

Know that I am with you and will keep you wherever you go, and will bring you back to this land; for I will not leave you until I have done what I have promised you.

Genesis 28:15

Pilgrim God, there is an exodus going on in my life: desert stretches, a vast land of question inside my heart your promises tumble and turn.

No pillar of cloud by day or fire by night that I can see.

My heart hurts at leaving loved ones and so much of the security I have known.

I try to give in to the stretching and the pain.

It is hard, God, and I want to be settled, secure, safe and sure.

And here I am, feeling so full of a pilgrim's fear and anxiety.

O God of the journey, lift me up, press me against your cheek.

Let your great love hold me and create a deep trust in me.

Then set me down.

God of the journey, take my hand in yours and guide me ever so gently across the new territory of my life.

Joyce Rupp

Words of Encouragement Along the Way

Throughout the challenging and difficult process of discernment, as a parish, we could have been tempted to focus only on those things that were readily visible. Such a surface-level approach might have been "easy," insofar as it would have limited the process to little more than determining the cost for repairing an old building and then arriving at a decision as to whether we should make that investment or simply close up shop. But the parish leadership body, along with myself, believed that we had before us an opportunity to go much deeper, making of this a spiritual journey for the people of God.

Many of the related themes and images of journey have already been reflected in preceding pages. At this point in our story, as a people neared the ultimate "last day" in the life of St. Joe's, I would like to recap the process by sharing a number of (dated) Harpe Melodies (bulletin excerpts) that served as an effective tool for encouraging all to go deeper into themselves as church, to look at the larger picture and allow this to be a spiritual journey unlike any other.

Harpe Melodies

10/22/00

....So, what now? We continue the process of deanery-wide planning, as part of the larger church. Within that process we have our own unique issues to address. Relative to the parish these issues are:

- Structural analysis of the church and recommended action.
- A thorough walk-through of our buildings and grounds, assessing needs as well as approximate costs for meeting those needs.
- Dialogue among leadership persons.
- Communication of findings, suggestions and progress to the larger parish.
- More dialogue and communication in terms of decision-making, planning, action and direction.
- The need to bear in mind that the deanery-wide planning process will likely impact the issues and, quite possibly, the decisions we will make as a parish.

Last week's column told of St. Francis of Assisi and his capacity to dream BIG. While this may be a difficult moment in the life of our parish, perhaps it yields a wonderful opportunity and invitation of **us** to dream....and to dream BIG within the very life of God. If I may paraphrase Psalm 33: Lord, let Your mercy be on us as we place our trust, our hopes, our dreams, and our lives in You. Peace and all that is good.
Fr. Dave

10/29/00

....Every person who walks through our doors is a blessing to all who gather together. At last night's meeting it was very clear that our parish is especially blest with people of deep faith, many gifts and an abundance of wisdom. And so, I say THANK YOU to the following: Joel and Linda Engel, Denise Ryan, Matt Kolkema, Linda Wilburn, Dan and Deb LeMire, Donna Ladd, Patty Siewert, Chris Nowak, Phil Konczyk, Tom LeMire, Jane Clingman-Scott, Dave Alexander, Ken Theisen, Don Liupakka and Joe Secord. I am confident that together, and with the Lord, you will faithfully lead us on the journey to wherever the Lord may be calling us as church.

Aside from praying our mission statement, I invite all to reflect on this little piece that Phil received via the internet:

> Life is like a river.
> Let it carry you,
> not knowing where it will take you,
> and you will journey to amazing places.
> Or, stay on the shore,
> knowing for sure where you will be,
> and you will go nowhere.

Paraphrasing Psalm 126: The Lord has done, and will continue to do great things for us. If that doesn't fill us with joy, then what will? Have a great week! Fr. Dave

11/5/00

....On Sunday afternoon, November 12, at 3:00 pm, there will be a parish town hall meeting in the family center. There are two purposes to this gathering:

- As part of the deanery-wide pastoral planning process, we will be looking at the demographics of our parish.
- A general discussion of the current situation regarding our old buildings and where to go with all of that.

On this Thirty-first Sunday of the year we are reminded (three times, no less) of the two greatest commandments by which we are to live: Loving God with all our heart, soul, mind and strength, AND loving our neighbor as our self. Hmmm. If we can keep these before us as we address the present and future needs of church, then perhaps the Lord will say to each of us "You are not far from the Kingdom of God." Peace and all that is good. Fr. Dave

11/12/00

Two unnamed widows serve as bookends to our readings this weekend. In Jewish society widows were understood as being in God's special care since they had nobody to defend their rights in a patriarchal society. Poor in spirit, they trusted deeply in God and were promised that they would inherit the kingdom of heaven. Utter dependence upon God, not self-righteousness, is rewarded. Once again, the world of worldly success and reward is turned upside-down....

....Where do WE spend our resources? In what, or in whom do WE place our trust? With all that is before us as a faith community, let us turn to these two unnamed widows for inspiration that is worthy of our imitation. And let us extend to them our gratitude for teaching us that God wants no more, and no less of us, than to offer others what we have to give. Together may we praise the Lord.... always. Fr. Dave

12/3/00

....We welcome and enter into a new church year. This short season of Advent (only 22 days this year) invites us once again to ground ourselves in faith, challenging each of us to ready ourselves to welcome the Lord who came to us in the Incarnation, the Lord who comes to us daily, the Lord who will come again in glory. Are you ready?

May we have the strength and faith to stand before the Son of Man when He comes. Peace....always. Fr. Dave

1/7/01

....This weekend we celebrate the Epiphany of the Lord. The readings remind us that our Light has come, not just for you and me, or for a particular group of people, but for ALL. So, what does this feast demand of us? It summons each one of us to be light for others, to be vehicles through which the Lord's justice and peace are further advanced. In short, the feast calls us to BE epiphanies (manifestations and / or reflections of the Lord's presence) to others, that every nation on earth will indeed adore the Lord. Toward this end, it is Herod whose instruction, "Go and search diligently for the child," might serve each one of us well, although for different motives than those of Herod. Why? Because if we ourselves are not always searching for and finding the Lord, it will be difficult and perhaps even impossible for us to BE epiphanies for others.

This weekend we also celebrated the wedding of Melissa Stout and Will Lee. Our prayerful support and best wishes be upon them as they begin their life together in Jesus Christ. May their marriage always be something of an epiphany for others.

And may the glory of the Lord shine upon us....always. Fr. Dave

1/14/01

...."Do whatever he tells you." These are Mary's words to the servants at the wedding in this Sunday's Gospel. They are also Mary's words to each one of us as we face any and all of life's challenges. But these words will have their greatest effect only when we are listening for the voice of the Lord and faithfully discerning His call in our lives.

Together may we listen for the Lord. Together may we do whatever He tells us....always. Fr. Dave

1/21/01

....With good timing, this weekend's liturgy presents us with Paul's wonderful analogy of the Christian community as a body with many parts. ONE body, ONE spirit. We're in it together, not for the good of any one member or any one group, but for the good of the WHOLE body.

Together may we embrace the Lord's words, which are spirit and life.

Fr. Dave

1/28/01

A person who worships regularly at St. Joseph made this statement in a recent phone conversation: "What keeps bringing us here is that you march to the beat of a different drum." I was then assured that this was intended as a compliment. Even so, the statement was very much before me for a couple of days. Is it possible that I am more different than I think myself to be? Could it be that this is simply one of those moments in which I find it difficult to accept and celebrate another's affirmation?

Each of us seem to wrestle with these questions at one point or another. The wrestling is itself good, for it often yields something of growth and insight. In my wrestling I came to the conclusion that I SHOULD be different. Further, I was reawakened to the belief that in the life of faith we are all called to march to the beat of a different drum.

With all that is facing this parish as well as the larger Catholic community of Muskegon, I am finding myself saying to people "Church is going to have to change, and it is going to look very different in the near future." This, too, is a tough one to wrestle with because it is going to demand a lot of letting go on the part of priests as well as the laity. It is going to demand letting go of the way in which we have experienced being church for many centuries... letting go and marching to a different beat.

Church is going to look very different in the not-too-distant future. The planning process in Muskegon as well as our unique situation here at St. Joseph, though difficult, should remind the entire Christian community that we are ALL called to march to the beat of a different drum. And this is good, for we are then simply imitating the example of Jesus Christ. Hmmm. There could be no greater compliment than to have this said to each one of us.

Have a great week, marching to the beat of a different drum. Fr. Dave

2/4/01

There's a phrase in West Africa called "deep talk." An older person (telling someone about a situation) will often use a parable, an axiom, and then add to the end of the axiom "Take that as deep talk." Meaning that you will never find the answer. You can continue to go down deeper and deeper....

Maya Angelou

As regards our parish's unique situation, the time for a decision is now upon us. Our leadership people met this past Tuesday evening to do some "deep talking" about where we find ourselves at this time. I will leave the details of that meeting to them, as they will be making a presentation to the parish next week. I do however, feel it necessary to print the following:

- For all parishioners, please make every effort to attend the Sunday Liturgy, February 11th. Immediately following the Liturgy, our leadership persons will address the congregation.
- Beginning with the weekend of February 17/18 we will, as a parish, gather for liturgy at Muskegon Catholic Central High School.

"Take that as deep talk," meaning that you will never find the answer. You can continue to go down deeper and deeper. Perhaps not in spite of, but rather **because** of our situation, we have a most unique opportunity to go deeper and deeper into the mystery of ourselves as individuals and as church. And if we do this together, I am confident that we will find ourselves going deeper and deeper into the ultimate Mystery of God, who loves us....who remains near and even within us.

Peace and blessings....always. Fr. Dave

2/11/01

It has been rather difficult to remain focused and upbeat as the parish comes to the point of making a decision about its very existence. During these past few months I have been grateful for the people with whom I walk, especially the people who are St. Joseph Parish. But there is one, a non-catholic and little more than a stranger, who touched and strengthened me in a personal encounter this past week....

Lori is about the same age as myself. Divorced, she is waiting tables at a local restaurant and struggling to carve out a living for herself and her children. The few times we have conversed, it was always clear that she is very proud of her children. They are her primary focus.

Last week, Lori approached me and told me that she thought I looked tired and stressed. I shared with her that the people of St. Joseph had found themselves at a most difficult moment in their journey. I then asked about her, for she herself looked tired and worn. Her eyes welled with tears as she spoke of how tough things have been in recent weeks for herself and her children, with the latest incident involving an old and failing furnace and carbon monoxide poisoning, which could have claimed their lives. But very quickly, Lori's expression changed and the tears gave way to a smile. She then said "But at least we're alive. Thank God."

We are alive and we are church, with or without our church building. Thank God! And thank God for the people who have helped to keep me / us focused, especially this people of St. Joseph, people like Lori and people like the prophet Jeremiah, who speaks to us this weekend: "Blessed is the one who trusts in the Lord, whose hope **is** the Lord."

God's peace and blessings be upon this community, today and always.

Fr. Dave

2/18/01

For several weeks we have begun our Liturgies by uniting our voices with the entire Catholic community of Muskegon, praying for this larger body of people who find themselves in a time of transition. Nobody is able to better appreciate this transition than the people of St. Joseph, as witnessed by the very difficult decision at which we arrived one week ago. We are in it. Transition is difficult because it involves change and letting go. It means that we will look different. But it also means this: We were church long before the establishment of a parish and the erection of a church building, and we will be church long after these have disappeared....because it is we who ARE the church.

Some will now begin connecting with other parish communities, while others will want and even need to remain together for a time. For those who will be gathering at Catholic Central in coming weeks, we need to watch and observe any number of things, some of them practical, but all of them important parts of our faith journey in this time of transition. Here are a few things to consider:

- Time will tell if we need to have two weekend Liturgies while at Catholic Central. We will simply wait and see.
- February 28th is Ash Wednesday. On that day we will gather for one Liturgy at 9:00 am in the parish's Family Center.
- Next Saturday 2/24, and Sunday 2/25, we will be celebrating baptisms, great reasons in themselves for us to gather together and do what we are called to do as church.

In many ways, our time of transition well seem awkward. Having no blueprint to follow will probably serve only to heighten our sense of awkwardness. But that's OK. For the people who gather together in coming weeks, if we do so with the same incredible faith that was witnessed here, one week ago, we will be fine. And in the end, our experience of transition might very well become a blessing, a cherished and even sacred moment in our faith journey.

May the Lord continue to bless each one of us....always. Fr. Dave

3/4/01

This past Tuesday the Liturgy presented us with a reading from the book of Sirach (35:1-12), which prepares us well for welcoming the season of Lent. The thrust of the entire reading is captured in verse 9: "Give to the Most High as he has given to you, generously, according to your means." Too often we think of Lent in terms of giving up something, only to take it back at Easter. And for some, their "giving" or "giving up" is accompanied by bitterness. Sirach reminds us that such grudging and forced giving is not giving at all. The words of Sirach hold true for the Christian: Our Lenten giving is to be accompanied by generosity, freedom, cheerful countenance and a spirit of joy. So, why is it that some still approach Lent with a spirit of negativity, as if it were a down time of drudgery? Perhaps there is some ungrounded fear that any true giving or giving up will somehow diminish us. On the contrary, our giving during Lent is what leads us to Easter. And our giving during this season, and always, simply makes us more Christ-like.

For the people of St. Joseph and for the larger Catholic Community of Muskegon, this year's Lenten journey holds great promise. As we move toward closing some doors and opening others, I invite all to be aware of the desert themes of Lent: Hungry and helpless, yet called to remain faithful, God's promise of a glorious future, Gathering at the water, Seeing more clearly, Having to die in order to rise, Facing the buffets and spitting. And when we arrive at Easter, be aware of its various themes: New beginning after hope was gone, The church earning the respect of the community, Standing up to the world, Growing and spreading out in the midst of a hostile world, Ministry and outreach, United (always) around what is central.

In the meantime, let us all be of good cheer, generosity and joy during this season of Lent....and always. Fr. Dave

3/11/01
Final bulletin column from St. Joe's

Let's begin with a couple of practical reminders:

- The weekend of March 17/18 will be our last gathering at Catholic Central.
- There will be no Saturday evening or Sunday morning Liturgy on the weekend of March 24/25.
- Our closing Liturgy will be celebrated in the church at 6:00 pm Sunday, March 25th. Given the occasion, we can expect a very large crowd. With that, I would strongly suggest that all parishioners who plan on attending make every effort to arrive well in advance.

Once again, the Lenten scriptures offer us nothing less than hope, promise and a new beginning. As we bring a parish to closure and move beyond the walls of St. Joseph, the readings for this Second Sunday of Lent present us with God's promise of a glorious future. We may be in a moment of sadness, but we have all that we need to sustain us as we go forth, namely, God's Word and God's promise. Therefore, we have, as the psalmist reminds us, **nothing** to fear.

May the Lord be our light and our salvation....always. Fr. Dave

The Buildings Come Down

The journey of the last several months in the life of St. Joe's proved to be profoundly spiritual. At the same time, ours was a painfully difficult path to trod. A people, a faith family for nearly 118 years, now found themselves as both observers and participants in an extraordinary event that can only be likened to an experience of dying and death. We were letting go of and losing much more than a building. As with the death of a family member or a close friend, we were losing something of our very selves.

The path that we had traveled together, in recent months, was emotionally demanding of all within the parish. But along the way, that path came to be dotted by a few key moments that served only to underscore all that had made St. Joe's, "the smallest of our parishes," one of the "best kept secrets in Muskegon." The first such moment was the parish's decision to seek closure, beautifully captured by the words of a senior parishioner, Joe, who stood before a people and said "It's time to let go." The second moment came with our closing Liturgy on March 25th, 2001. Difficult? Yes. But what a witness this was to a people's faith, to their deep trust in our benevolent God. And now, the people of St. Joe's were fast approaching a third key moment in the final stage of our journey as a parish, the razing of all the buildings at 1224 Fifth Street, Muskegon.

By mid June, our buildings had been vacated, stripped and readied for demolition. All that remained was a ten-day period of waiting, which for many of us became something of a vigil. It was as if we had been called upon to quietly and prayerfully keep watch as a loved one lay dying. But even during that period of waiting, of keeping vigil for an old friend, God continued to provide us with light, inspiration and hope.

Just a couple days before the buildings were to be demolished, I walked one last time through the old rectory. It had served as the residence for many priests, and for nearly four years it was the place that I called "home." Upstairs, I returned once more to the room that I used as both a study and a place of prayer. Within that particular room I had found comfort, solitude and a deep sense of security, for just a few yards beyond the side window sat the old church. As I took one final look from the same window, my attention was instantly drawn to the church's side door. It appeared that we had had a visitor, apparently a child, who with the use of sidewalk chalk felt compelled to leave these words on the door: **I LOVE YOU, GOD**.

On Sunday evening, July 1st, I returned to the site of St. Joe's. With the day of demolition just hours away, I wanted to make sure that the buildings were secure. But of greater importance was my personal need for one last quiet, prayerful and reflective walk around the grounds. As I made that walk, I noticed a Muskegon Police patrol car parked along the side street, very near to the church. Approaching the vehicle, it appeared as if the officer behind the wheel had also come to pay one last visit, to quietly reflect. I introduced myself to the officer, whom I had never met, and then spoke these words of request: "Our old friend, here, is coming down tomorrow. I don't anticipate any problems between now and then, but I would like to make sure that the buildings and grounds are not violated in any way. And so, I am asking if it would be possible to pay special attention to this place tonight." The officer, appearing to sense my feelings as well as those of my parishioners, responded: "I'm on duty all night. And I will gladly keep a close eye on your old friend." These words provided assurance and comfort for myself as well as for many others on this particular evening. For this we are grateful.

Monday, July 2nd, came all too quickly, and our final moment was now upon us. Arriving on site at 7:30 am, to open the parking lot for parishioners and other likely observers, I was not at all surprised to find Rita Sander there to greet me. Her entire life remembered and celebrated at St. Joe's, Rita had also served for many years as the parish's sacristan. In that capacity, it was her task to make sure that things were prepared and "ready to go" when it came time for weekend Liturgy, and that everything was cleaned-up and put away afterward. It was only fitting, therefore, that Rita, affectionately referred to as "the church mother," would be waiting to greet me on this day.

At 8:00 am the work of demolition began. Two men, with the use of one very large piece of machinery, proceeded to bring our buildings down. The workers performed their task with great care, precision and skill. But at the same time, they seemed keenly aware of the faces surrounding them, many of which were awash with tears. The first building to come down was the parish family center. Next were the garages, followed by the church. And finally, the rectory came down. By 4:30 pm three quarters of a city block, a place of life and beauty for well over one hundred years, had been reduced to a field of debris. All that remained, which nothing could ever destroy, were countless memories.

Despite the noise from machinery and the toppling of buildings, the day of demolition was marked by prolonged periods of notable silence among those who were gathered. Parishioners, neighbors and total strangers stood together. While some folks had come simply to watch, most were there to remember and to further let go of something precious. Throughout the day, and especially within the silence, one could not fail to hear the voices of care, support and sympathy for the people of St. Joe's. In and around this difficult moment, we knew that God remained very near.

Although few words were said throughout the day of July 2nd, the images and events of that day speak volumes. What follows are but some of the images and encounters which, for me, will forever speak of the depth of a shared experience that cannot be captured by words alone.

Within minutes of my arrival, on the day of demolition, I saw Eddie Jenkins approaching on foot. Tall, black, unassuming and well into his seventies, he and his wife, Nancy, were both neighbors and long-time parishioners of St. Joe's. The parish was the spiritual home in which they raised a large and beautiful family. They, along with many others, were now losing that home. While Nancy was not yet ready to watch the buildings come down, Eddie was determined to be there from the start. As I greeted him with an embrace, he spoke these words: "Look the other way, Father, unless you want to see an old man cry." I assured Eddie that he was not alone. And I thanked him, for his were the tears and the feelings of many.

From a large number of our immediate neighbors, many of them non-Catholic, there came words of sadness and sympathy. But there were also expressions of their own sense of personal loss, which were captured in the words of a young, single mother: "This neighborhood has known a lot of trouble and sadness over the years. But that old building has always been there. And now, it's going to disappear." Once again, I was reminded that the loss of St. Joe's was not ours alone. It would be a loss that reached far beyond the people of the parish.

A most powerful image, and now a memory, presented itself as the church was being demolished. With much of it already in ruins, all that remained standing was the belfry and, directly beneath it, what had been the main entrance to the church. The intent of the workers was to bring the tower down by simply pushing it over. A relatively easy task, one might think. However, it appeared that there was something present within the structure itself that was not yet ready to give up and let go, as if the tower was deliberately resisting the workers' force against it. After a couple of failed attempts, accompanied by expressions of bewilderment, the workers reengaged in what was beginning to resemble a tug of war. Repositioning themselves and their equipment, they delivered one great push. Finally, the belfry succumbed. Teetering for just a brief moment, the tower then collapsed in its entirety.

By 5:00 pm, many of the observers had dispersed, going their separate ways. What had been the site of St. Joseph Catholic Church now lay in ruins, and workers busied themselves with cordoning off the lot and carefully piling bricks in such a way that they would be accessible to parishioners as souvenirs. As I watched and further reflected on this final act in what my brother, Gary, had previously referred to as the parish's "swan song," a gentleman named Bob Beecham approached me. He was no stranger to the parish, for Bob and his siblings had been raised in the faith at St. Joe's. Having long since moved away from Muskegon and into another parish, he had returned in recent days as part of his job, which was to coordinate the demolition of our buildings. As we stood chatting, Bob drew my attention to one of the workers, the one who operated the machinery that had leveled everything around us. He then went on to say that a transformation of sorts had just occurred within that same young man. It seems that until a few days prior to demolition, Jim was wearing his hair in dreadlocks, had grown a long beard and mustache, and was sporting a couple of earrings. Suddenly, all of these were gone. Standing before us was a rather clean-cut and handsome individual. Bob then told me that he had asked Jim "Why the change?', to which Jim responded "I've never torn down a church, and I just think I need to clean up for this job."

I went out of my way to speak to Jim, and to thank him. I then informed him that I was aware of his recent "transformation," and that I was deeply touched by his sensitivity. He smiled and then left me with these parting words: "You know, this is holy ground." "Amen" could be the only response to such a statement. Yes, this **is** holy ground.

It is Finished, Or Is It?

> After that, Jesus, realizing that everything was now finished, said to fulfill the Scripture, "I am thirsty." There was a jar there, full of common wine. They stuck a sponge soaked in this wine on some hyssop and raised it to his lips. When Jesus took the wine, he said "Now it is finished." Then he bowed his head, and delivered over his spirit.
>
> John 19:28-30

Leaving the patch of holy ground that was the site of St. Joe's, I returned to the home of Matt and Tanya Kolkema and their beautiful children, Derek and Colin. This wonderful family had opened their door to my cat, Jeter, and me since that day in mid June when everything was permanently shut down at St. Joseph. As I drove along, nearing the end of an emotionally draining day, I found myself saying aloud the words of Jesus: "Now it is finished." By the time I arrived at the house, having repeatedly spoken those words, I became aware that I was smiling. And I began to ask myself "Is it really finished?" For Jesus' passion and death were not an end. No, the story would continue. He would live, and He would remain with us. Could it be that His passion, death and resurrection were, in some tangible fashion, being played out in the very real experience of a people? We knew something of suffering and death. With our parish closed and the buildings demolished, it would certainly appear that now it was finished. But our story, our faith would continue. These would live on, and they would remain for others.

July 2nd, 2001, came to close on a rather bittersweet note. I was at home with the Kolkemas'. Late in the evening I sat and visited with Derek, who was then fourteen years of age. For a time we simply reflected on the events of the day, and we shared some of our memories of the last few years in the life of a parish. At one point there was a brief silence, and Derek's face became somewhat gaunt. Then, looking directly into my eyes, he spoke these words: "Father, now we won't have a good priest in Muskegon." Through my tears, I tried to assure him that I believe there are some good priests in Muskegon, and that what they might need is to learn how to simply let go and allow themselves to enjoy what should be an incredible journey with God's people. Again, there was a moment of silence. Then, with a tender embrace, we said our good nights.

Chapter Three
MOVING BEYOND

Chapters 1 and 2 of our story had to do with possibilities, of what church **can** be if we but allow ourselves to think and dream big, especially with regard to matters of faith, and dare to take some risks with God and with one another. For the people of St. Joe's, this was clearly witnessed in the following:

- The welcoming of all – be they parishioner, family member, friend, or complete stranger.
- Sharing and honoring the many gifts and talents within the faith community.
- Stretching and challenging one another along the way.
- Claiming and/or reclaiming our faith.
- Appreciating and celebrating the fact that we journey together as a people, not as individuals who just happen to share a common space at prayer and worship.
- Knowing that we are always in the process of being shaped, changed, renewed and recreated as we continue to grow into the image of the divine.
- Experiencing Liturgy as the center, the heart of it all.

Now we come to Chapter 3 in our story, which has to do with **reality** and the lived experience of many within the church today. Before continuing, however, it might prove helpful to present something of a backdrop, against which the remainder of our story is written. Hopefully, this will provide the reader with a snapshot of the present reality within Catholicism.

First, I wish to share a few words regarding matters of a spiritual nature, that deepest level of one's faith and being. As the people of St. Joseph were blest to have known something of the possibilities, of what **can** be, far too many Catholics are witnessing quite the opposite, the sad and even painful reality of what IS. This too is being echoed, through a growing chorus of voices, in statements that have become all too common and which continue to fall upon the deaf ears of our church leaders. The following are but a few of the more familiar expressions of that which many are experiencing within the church today:

- We are not being fed spiritually.
- Preaching and presiding, by priests and bishops alike, is poor and lacking.
- We come to the table hungry, only to leave starving or, at the very least, angry.
- They (the bishops) refuse to listen to us.
- We (the laity) have no voice.
- We (the laity) feel as though we are nothing more than a source of income.
- I am (or "we are") open to looking at other Christian traditions for spiritual nourishment.

Secondly, a few words are called for at the external level of religion, the particular avenue by which one nurtures and practices his/her faith (e.g., Catholicism, Judaism, Islam, etc., etc.). For many within the Catholic community, the present reality is given witness in these all-too-familiar expressions:

- Church has become "big business."
- Church governance needs to be decentralized, with bishops and bishops' conferences enjoying greater autonomy.
- Papal ministry has become a dictatorship.
- The Second Vatican Council's vision of Episcopal collegiality has all but disappeared during the papacy of John Paul II.
- Rules have become the religion.
- The magisterium (the pope and the college of bishops) is out of touch with the people of God.
- The laity must be given a greater voice in such matters as church governance and the selection of bishops.
- Growing is the number of persons who are minimally connected to the church, or who are now identifying themselves as "former Catholics."

- Offering band-aid responses only, the magisterium is failing to adequately address the growing crisis of a rapidly shrinking supply of priests.
- The celibate requirement should be lifted, and priests be allowed to marry.
- We should be ordaining women for priestly ministry, or at least discussing the matter openly.
- There is no room, no forum for dialogue within our church.
- We (the laity) have no voice in Catholicism.
- The pope and the bishops fail to hear us.

Thirdly, there is the clergy sex abuse scandal. And, as if adding insult to injury, there is the scandal of cover-up and conspiracy on the part of our bishops and many of their predecessors, with Rome very likely to have been involved all along. Now, having brewed for decades, the volcano that erupted at the end of 2001 is sure to be spewing forth lava for years to come.

Fourthly, I invite the reader to reflect once again on this question, which appeared in the early pages of our story:

How many of us have a brother, sister, mother, father, son, daughter or friend who are minimally connected to the Catholic Church, or who no longer consider themselves to be "Catholic?"

Finally, as a way of underscoring the preceding paragraphs, I would like to take this opportunity to articulate the present reality by giving the reader the "short version." In the course of my writing, a great deal of time has been spent between two vastly different communities in North Texas: Urban Dallas and tiny, rural Charlie, some twenty miles from Wichita Falls. Although they are worlds apart in so many ways, both serve as reminders that when we cut through the complexities of life, God's people remain very much the same. This holds true with regard to matters of faith and religion, which all-too-often become extremely complex. And whether it's Dallas or Charlie, Texas, many are the times in recent months that I have heard the following expressions which, in wonderfully uncomplicated fashion, capture what IS the sad reality for a growing number of Catholics:

- "We've got the fox guarding the chickens."
- "When you sleep with dogs, you're gonna get fleas" (a direct reference to the bishops and the clergy sex abuse scandal).

It is against this backdrop, that of sadness and discontent, that our story continues. As a reminder, however, while part III is primarily the sharing of my personal journey during the past few years, it reflects the experience of many within the household of Catholicism. Now three years since the closing of St. Joe's, it saddens me to know that a number of my former parishioners remain very much unsettled. Some have not yet connected with another parish. Others are finding spiritual nourishment by gathering, weekly, in the homes of one another, there breaking open and prayerfully sharing the Sunday scriptures among themselves. For a small handful of individuals and/or families, spiritual needs and hungers are now being satisfied in a non-Catholic religious tradition. Here, again, is the reality: God's people are hungry, and our church leaders (for the most part) continue to offer little more than swill. With the laity having no voice and finding that there is no room for dialogue within the present institutional structure, many Catholics are now feeling as though they have been pushed to the edge. At the same time, growing is the number of the faithful who are simply leaving Catholicism, going elsewhere in search of spiritual nourishment.

Because our faith, our stories and our very lives are so intertwined, one cannot share their own journey in isolation from the whole, that is, apart from our journey as a people. Thus, the most effective way for sharing my personal journey is to do so in the company of others who, together, travel the path of faith. And so, the bulk of this portion of our story, MOVING BEYOND, will consist of a chronology of letters, some of which might require further commentary and/or clarification, but all of which are reflective of the pain, sadness and frustration that continues to burden the hearts of many.

Letters -- 2001

Spring, 2001

Dear Bishop Rose,

We are contacting you regarding an issue that we are finding quite troubling. We are confident you will give consideration to our concerns as they are shared by several Catholic families we have spoken with in Muskegon. We are going through a difficult and challenging process in the Deanery, and you have an opportunity to provide leadership. Quite simply put, there seems to be a hunger that is not being fed very well in Muskegon. So much so that there are actually families considering leaving the Catholic faith for something else. These are good spiritual families that participate with their finances and their talents. Families whose children should not only continue to be brought up in the faith but who should be inspired to attend our Catholic schools.

Our family is one of many unfortunate casualties of the closing of St. Joseph. I understand business decisions must be made and fiscal responsibilities are important, but so are the sheep. The closing of the parish was much like a death and the good people of St. Joseph's truly needed an opportunity to mourn, and an opportunity to see the sun rise with real options and real hope. You referred to the closing as a "blueprint." You were partly correct in that we made a responsible decision on what to do with the building. The people part of the parish (perhaps one of the most lively parishes in town), however, senses that we have been left to scatter and make the best of the circumstances with no follow-up. Businesses that allow their clients to wander typically don't do very well and I have come to expect more inclusiveness, guidance and concern from my church. I am still hopeful that concern will surface. Father Mike at Our Lady of Grace has been very welcoming and kind to the people of St. Joseph's. Please don't misunderstand us. This isn't about a pity party or feeling that we have gotten a bad shake. It goes deeper and brings us back to the point of being fed.

When the people of our parish recognized it was time to close the doors and not the heart of St. Joseph's, we asked you to try very hard to allow Father David Harpe to somehow remain in Muskegon. We understand the Appointment Committee makes these decisions, but we also believe that Muskegon needs Father Dave and your support of that to the committee would be valuable input indeed. On a personal note, Father Dave encouraged our daughter to be an Altar girl, our son to be a Lector, my wife to assist the Music Director and Eucharistic Ministry and myself to be involved in the choir and lectoring. He draws that passion from those he touches, but I am sure I don't need to tell you that.

We have been blessed by God's grace in finding excellent churches and pastors in the places we have lived. In Fort Wayne, IN, Monsignor John Seultzer at St. Charles Borromeo, in Holland, MI, Father Bill Duncan at Our Lady of the Lake, and after visiting several parishes in Muskegon, we found Father Dave at St. Joseph's.

At a recent Deanery meeting, it came to our attention that conversation was reduced to raised voices, vulgarity and divisive words. Hardly what one would expect from a committee formed to bring us all together. We know Father Dave serves and welcomes all Catholics and believes that given the opportunity, he would remain in Muskegon. And so we are appealing to your good judgment to help find a place for him here. We are hopeful all the other priests of the Deanery would support him in staying.

Thank you for taking the time to read this letter, and we look forward to your response. We know that Father David Harpe will serve the Catholic Church and his parishioners well wherever he is. We believe given the socio-economic and ethnic diversity of Muskegon, that all would be best served to have him celebrate here. We hope you agree.

Most sincerely and humbly,
Darrell, Marcia, Nick and Rachel Senior

To date, the Senior family have received no response to their letter. Numerous phone calls to the office of the Bishop have likewise gone unanswered.

At or about the same time that the preceding letter was written, I had an occasion to visit with the bishop. In touch with my own feelings of sadness and loss at the closing of St. Joe's, feelings that were shared by many, I recall saying to the bishop "You have a wonderful opportunity before you." With a people hurting and mourning, I suggested to the bishop "As a pastor, please consider doing a general mailing to my former parishioners. It need not be lengthy, but just to assure them that you remain with them in prayer during their time of loss and transition." I went so far as to offer to cover the costs, for such a mailing, out of my own pocket.

There was no general mailing. This only led myself and others to further believe that in the end, at the diocesan level, the closing of St. Joe's was little more than a matter of "business": A problem emerged, and it was effectively addressed. Now it was time to get back to business as usual.

August 7, 2001

Dear Dave,

This is the first chance I have had to answer your letter of May 20. By now I presume that you have moved into St. Bernard's and have had a month or so of experience with your two new flocks. Two quite different flocks, I would think. St. Bernard's has not had a young American priest since Tom Boufford, and St. John Cantius since Tom Bolster. I believe they will appreciate your energy and prayerful service. It may take each of them a while to get used to the change.

Among other things, I hope you will be able to continue the work that Father Niedzwiecki had begun with AA. The need seemed to be very great. And Tom Boufford was much involved with Manistee Catholic. I don't know whether that is still possible and desirable or not. In any case there certainly are the rural poor in the area; they are mostly "invisible', but they are there. I know you have a special sensitivity to them.

I appreciate your letter, Dave. I too have had great hopes for the Muskegon planning process. Muskegon is in many ways our most homogeneous deanery. I really have hoped that it could provide the model for several other deaneries. The Mass last fall was encouraging, as were the comments of the priests and people afterwards. Since then, however, as the time has come for making some hard, concrete decisions, it is obvious that the process has become much more painful.

Our people are no different from everybody else. Most do not like change, especially change which is taking them into the unknown and change which seems to involve some loss of local control. And, as we learned after the Council, change in something affecting people's religious life is perhaps the hardest of all, because it affects people at such a deep level. Change at that level comes about only by conversion.

I think that many people in Muskegon became alarmed at the talk of a possible "one parish for all" arrangement. That is a very radical re-alignment, with no obvious precedents anywhere that we know of. I believe that proposal finally brought people into the discussion who had paid little or no attention to what was going on earlier. After all it is only a relatively small group from the thousands of Catholics in Muskegon who have gone through the preparatory stages of the process. Many, if not most, of the people have had only a vague idea of what was going on.

As of now, from what I hear, the immediate decisions will be much more modest. I hope that there will be at least the beginnings of some genuine and serious collaboration between parishes and among all the parishes. We will see. Planning can be a very useful tool, but salvation comes through God's grace. And that we cannot plan.

I appreciate the fact that you would have been willing to remain in Muskegon, but I hope you realize that that was simply not possible. We needed you right where you are now. And I think, after the painful experience of closing the parish of your first pastorate, it is important to have a breather from that kind of tension and pain. You would not have had any kind of breather in Muskegon. I hope you will find it in the north. You deserve to simply enjoy being a pastor for a while. I hope you can at St. Bernard and St. John.

I will be at St. Stan's in Ludington for their centennial next Sunday afternoon. If you are able to attend, perhaps we will have the chance to talk for a few moments. I will be interested in your first impressions of the north.

With every best wish, I am
Sincerely in our Lord,

(Most Rev.) Robert J. Rose
Bishop of Grand Rapids

19 August 2001

Dear Bishop Rose,

Greetings. This is one of those rare occasions in which I begin a letter with an apology of sorts: I am sorry if what follows sounds volcanic in tone, but there is no better adjective for describing what I am feeling and experiencing at this moment. With that said, I will proceed.

I appreciate your letter of response (8/7/01) to that of mine (5/20/01). Near the end of your letter, responding to my desire to have remained in Muskegon, you made this statement:

> "We needed you right where you are now. And I think, after the painful experience of closing the parish of your first pastorate, it is important to have a breather from that kind of tension and pain."

Your words convey a genuine concern, for which I am grateful. However, the reality is that my present experience is fast shaping up to be anything but a "breather." On the contrary, I feel as though I am living one of the worst of nightmares. On virtually every front, the parishes of St. John Cantius and St. Bernard have suffered near unimaginable neglect, with St. Bernard being, by far, worse off than St. John Cantius. While aware of Fr. Tom's illness during the past year and a half, it would be apparent to even a total stranger that such neglect long preceded his sickness. At a glance, I would go so far as to suggest that things were in pretty poor condition (at least at St. Bernard) when Fr. Tom himself arrived as the new pastor.

At this moment I am feeling angry, frustrated and overwhelmed by this situation. Quite frankly, I am also feeling as though I have been blindsided by all of this, insofar as our priest appointments committee had to know that things were not good, here. And yet none of this was communicated to me along the way.

In an attempt to help you grasp something of the seriousness of this situation, let me begin to paint a picture. And at this point, begin is all I can do.

I am living in what can best be described as squalor. I do not eat meals at St. Bernard because I refuse to prepare food in such an unhealthy environment. It is so bad that I actually fear for the well being of my cat. It is likely that I will be moving into the rectory at St. John Cantius. However, that house is in need of attention as well. It seems that some major upgrading and improvements were begun a few years ago, but I'm told that Fr. Tom abruptly halted the project when he first became ill. Why, I do not know. In the meantime, I am hoping to soon host an open house for the parishioners of St. Bernard, inviting them to walk through this rectory and then ask themselves if they would live here.

The visible problems are not limited to the rectory at St. Bernard, but are reflective of the entire plant, which is a glaring monument to piecemeal and the very creative uses for duct tape and silicone caulk. And what is visible is further reflective of the internal neglect at both St. Bernard and St. John Cantius parishes, which at a glance looks like this:

- Presiding at Liturgy is something of a chore, so lacking are the key elements of music and congregational involvement. It has been nearly forty years since the Council called for "full, active and conscious participation" in the Liturgy. But the experience here, especially at St. Bernard, is more like "empty, inactive and unconscious presence at Fr.'s Mass."
- There is no Liturgy Commission, Finance or Pastoral Advisory Council at either parish. When such groups were meeting, it was not on a regular basis. And at both parishes I

have been told that it has been at least a couple of years since the last meeting of any one of these leadership bodies.

- There is no such person as a sacristan at St. Bernard. Fr. does it all – from opening buildings to setting up, cleaning up and locking up afterward.
- Prior to my arrival, there were no collection counters at St. Bernard. On the Monday after my first weekend here, the bookkeeper was surprised to find that the weekend collection had not been counted and made ready for deposit. She then informed me that it was Fr. Tom's practice to privately count the collection himself. Upon catching my breath, I informed the bookkeeper that such a practice would end immediately. We have yet another band-aid approach in place for the moment, but at least the pastor is no longer in the position of handling the collection.
- I've asked a few individuals about the R.C.I.A. Their response was something like 'R.C.I. what?"
- Unique to St. Bernard is the parish's involvement in and support of Manistee Catholic Central High School. It was around this issue alone that the honeymoon ended a few weeks ago during an open meeting of interested parishioners who were gathered for the purpose of expressing needs and concerns of the parish, as a way of helping to pull things back together. The school was the hot topic right from the start, around which the following points emerged:
  o Manistee Catholic Central had a total of 265 students (grades K-12) for the 2000 – 2001 school year. That number is expected to remain much the same for the coming school year.
  o St. Bernard currently has sixteen students enrolled at MCC.
  o We are unable to document any financial support of the parish by nearly half of the parents of the sixteen students St. Bernard has enrolled at MCC.
  o For the coming school year, St. Bernard has been assessed $46,900.00, which amounts to nearly half of the parish's yearly Sunday income.

- o Over and above the school assessment, the parish pays another $10,000.00 for the local Dial-A-Ride service to transport St. Bernard children to and from school. With that, we are now over half of the parish's yearly Sunday income.
- o In the very near future MCC will be faced with having to make a high-dollar investment in building repairs and improvements.

Concern was loudly voiced at the parish's open meeting that St. Bernard's relationship with MCC must change. With a largely retired congregation and only sixteen students enrolled at MCC at a cost of nearly $60,000.00, I can only agree that there is something radically wrong with this picture, and that it must change. It's as if we are sailing the Titanic.

Relative to the school, I asked the parish bookkeeper why Fr. Tom chose not to broach this matter especially when it is obvious that the parish has been hemorrhaging money into MCC. The response was that "he didn't want to upset anybody." This much is very clear to me: In his silence, Fr. Tom has upset many.

On a personal note, Bishop, if I could choose between this situation and that of the closing of St. Joseph Parish, Muskegon, my choice, without hesitation, would be the latter. Believe me, this is no "breather." And if I may be so bold, it occurs to me that the picture I have begun to paint is not simply that of two broken parishes, but a much larger church that is broken and in serious need of repair.

Thank you, Bishop, for at least allowing me this opportunity to ventilate. I will look forward to further discussion of all the above.

Sincerely yours,
Rev. David L. Harpe

Although harsh in its tone, the preceding letter is not a reflection of the people who make up the parishes of St. Bernard and St. John Cantius. On the contrary, they are persons of deep faith who, like so many others, are hungering for spiritual nourishment. Unfortunately, within the present leadership structure of the church they, again like many others, have themselves become casualties of a rapidly crumbling system.

17 October 2001

Dear Bishop Rose,

Greetings. It is with a heavy heart and clear conscience that I write this letter, sharing with you that which I preached to the people of St. John Cantius and St. Bernard parishes on the Twenty-Eighth Sunday of the Year.

First, I stated directly: "You might very well see this priest walk away from priestly ministry." I then stated that we have a problem. And this problem is not mine. Nor does it belong to any one person, parish, or diocese. Rather, it is a problem that belongs to the whole church. The problem – the shrinking number of priests and the subsequent stretching of active and retired clergy in what appears to be a failing effort to meet the needs of the Catholic community.

Second, I proceeded to speak to two congregations about leprosy. I suppose I could have played it safe, but I didn't. Instead, I spoke metaphorically about the leprosy that seems to be permeating the very leadership of our church, namely, fear, denial and desperation. With so much energy spent on plugging holes and frantically filling in the cracks, I feel as though the priest is being called upon to become more of a businessman and less of a doctor of souls or bearer of the mysteries. And as a result, many other leprosies will go unnoticed, untouched, unhealed.

Finally, in reference to Second Timothy, I told two congregations that I suspect most, if called upon to do so, would endure hardship for the sake of the gospel. But must any one of us endure hardship that is unnecessary? Especially at the expense of one's emotional, mental, physical and spiritual well-being? As I said to the people of St. John Cantius and St. Bernard: "Called to a life of service is one thing. But no one is called to a life of abuse."

In closing, Bishop, this is the most difficult letter I have ever written. I hope that you are not personally offended by my words, and that you and I will have an opportunity to further reflect on all of this. Please know, however, that I will be away for nearly three weeks, beginning November 4th. Thank you.

Yours in Christ,
Fr. David L. Harpe

HARPE MELODIES

10/21/01
Last week I spoke very candidly about an issue that cries out for open and honest dialogue within the larger Catholic Church, the issue being the shrinking number of clergy. The two primary concerns that I believe were conveyed are:

- The continued stretching of our current supply of priests, both active and retired, and how this compromises priest's effectiveness as well as their emotional, mental, spiritual and physical well-being, and
- The compromising of the needs of the people in the pews who ARE the church. One profoundly effects the other. And as stated one week ago, my greatest fear is that as this situation only worsens it is the people of God, as a whole, who will pay the greater price.

Following last week's homily, I was anticipating the possibility of a backlash of complaints. As of this moment, nothing of that nature has occurred. But that doesn't mean there are no such complainers out there. However, I was pleasantly stunned by the immediate positive response from any number of individuals following each of last week's liturgies, regarding both the problem, which belongs to the larger church, as well as the personal decision with which I am wrestling, namely, do I stay or walk away from priesthood. To all who have spoken or written words of support and encouragement, THANK YOU from the bottom of my heart. And to all, whether we are in agreement or not, this much seems most certain: This Catholic Church, of which each of us are a part, must sooner or later confront some critical issues. And the voice of the whole church must be engaged in that dialogue.

Finally, as was also stated last week, I have written all of this to Bishop Robert Rose. The letter was placed in the mail on Thursday, and I am confident that a response will come rather quickly. As we pray for one another and for the well-being of the worldwide Catholic Church, perhaps we could make Paul's words our own: "Beloved: Remain faithful to what you have learned and believed, because you know from whom you learned it, and that from infancy you have known the sacred Scriptures, which are capable of giving you wisdom for salvation through faith in Christ Jesus." (II Timothy 3:14-15)

May the Lord, who is our help, bless us....always. Fr. Dave

15 December 2001

Dear family and friends,

Greetings and peace in Christ. With a heavy heart I am writing this letter to inform you of my decision to take an extended leave of absence from priestly duty, effective as of December 5th. Many issues have contributed to this very difficult decision, none of which have to do with any person or persons within either of the two parishes I am currently serving. So, how is it that I have arrived at such a decision?

At a personal level there has been a growing fatigue, which has compromised my effectiveness as well as my health. To state it directly, I am experiencing what millions of persons wrestle with, namely, some degree of depression.

At a broader level I continue to wrestle with many of the issues that are facing the larger Catholic Church. As you know, I have been (and will continue to be) vocal about such issues, for example, the shrinking number or priests and the subsequent stretching of active clergy to breaking point, all for the purpose of maintaining the church structure, as we know it. And until the larger church addresses such issues, that structure will continue to crumble at a much faster pace than the current church leaders' ability to plug the holes.

My decision to take a leave of absence will, no doubt, present some inconveniences while at the same time underscoring this problem which belongs not to me, but to the larger church: finding yet another priest to "fill in the hole." And while I feel compelled to apologize for any inconveniences, especially as they affect my parishioners, I believe it is the larger church leadership that owes all of us an apology.

Perhaps there is something good about the timing of my decision. We find ourselves in this wonderful season of Advent, a time of remembering Christ's first coming among us, a time for anticipating Christ's second coming, each of which are to help us better welcome Christ's coming among us in all our daily living. If I may play on that word "Advent," perhaps this is a time of "coming to" for each of us: Coming to a greater awareness of who we are, coming to a deeper life in Christ, coming to a greater sense of responsibility for being church and living out our faith. This is the stuff of Advent, without which Christmas is just another passing day.

As I prepare for departure for a time of reflection and discernment, I ask for your prayerful support not only for myself, but also for the larger church family to which we belong. May Advent lead each of us to a Christmas filled with joy, blessings, family, friends, good health, wholeness and happiness. May Christmas be anything but "just another passing day."

Prayerfully, and with love,
Fr. Dave

19 December 2001

Dear Pope John Paul,

This is the first and perhaps only time that I will send Christmas greetings to you. Enclosed, you will find what I have written and mailed to hundreds of households in recent weeks (letter, dated 12/15/01). I invite you to read and reflect on it as well.

As I entertained thoughts of writing this particular letter, it was with a desire to give further expression and clarity to this rather sad moment in my life. However, this is a church whose leadership is not prone to dialogue, especially with the laity. Convinced, therefore, that this letter will only fall on deaf ears as well as hearts and minds that are closed, I will not engage in an exercise of such futility. I will, however, leave you with a couple of parting thoughts and/or observations: If Rome is to be the heart of Catholicism, then it would appear that the church has suffered a massive infarction. Out of time, out of touch and out of gas, the current leadership of the church, especially in Rome, stands at the threshold of spiritual bankruptcy.

As I enter this period of personal reflection and discernment, I will find comfort in the support of God's people and in God's promise "I will not abandon you." And I will pray that God send us another John XXIII or Paul VI…. Soon.

With all due respect,
Rev. David L. Harpe

Letters – 2002

1/06/02
Dear Father Dave,

I thought of you again today, this time during St. Michael's Mass, as the visiting priest was sprinkling the congregation with holy water via the usual method. I always preferred the pine branch you used. It just seemed more natural and more connected to the people. It is only one of the many, many things Ron and I miss about having you as our priest. We still have not found a church family to compare with St. Joe's, although we attend Our Lady of Grace more frequently than the others. Sadly, there are not many options in our Muskegon area. Myself, I feel like I am wandering in a spiritual desert of Catholic churches, occasionally being refreshed by an oasis like a CTA conference.

Thank you for the letter informing us of your decision to take an extended leave of absence from the priesthood. Our thoughts and prayers are with you during this difficult period of your life. What a hard struggle it must be for a person who has devoted his life to the church. It is not an ordinary career. As I thought about your dilemma, it occurred to me that being a priest is in many ways more like a marriage than a job. And the only thing (in my experience) that I can equate to your struggle would be separating from a marriage partner of many years and considering a divorce. A marriage that is not working because 1) you have both grown and evolved in different directions, or 2) one of you has not evolved at all (and I believe the latter fits this old dinosaur of a church).

In truth, I wish I could motivate you to continue the priesthood, by saying: "Hang in there, it is people like you who will help change this church," but I still have not jumped in myself and become a Catholic. It seems like whenever I start contemplating it, Cardinal Ratzinger, the Congregation for the Doctrine of the Faith and its committees, or the pope comes out with some archaic statement that I know I could never identify with or want to be a part of. I have thought long and hard about trying to change this church from within, but I seriously doubt whether there will be any major changes in my lifetime. So perhaps, I am not the person to pump you up to remain a priest. However, you know you are an excellent writer. Have you given any thought to documenting and publishing your struggle? At the very least it would be cathartic. At best, it could add to the movement to revitalize this church by making more people aware of the growing problems in the Catholic Church.

Please keep in touch and let us know what you are doing. Stop in and see us when you are in Muskegon.

Take care,
Karen Pavlich

January 12, 2002

Dear Dave,

Thanks so much for thinking of me and sending your letter. While I have to say that I was stunned, I honestly have to say that I'm not surprised. I've admired you as someone in touch with your feelings, with your heart, with your soul. You have always seemed like a person filled with the Spirit. My own theology says that one who is filled with the Spirit is constantly challenged (and challenging) to think outside the box, to ask why, to look for possibilities. I think that's pretty much the definition of a prophet. And prophets don't fit too well in the institution these days.

When I was struggling with my own process of discernment, I had the opportunity to attend the annual Corpus convention. What an eye-opener that was! I also plugged myself into a local network of non-canonical priests. I say that because they are still very much active in a sacramental ministry that focuses on weddings, funerals and Eucharist for the 70% of Catholics and others who live on the fringe of the institution. They have been a wonderful fraternity of support, guidance and encouragement for me. I have to say that I feel more akin in a brotherhood sense to the married priests here than I do to the single (not necessarily celibate) priests, not one of whom reached out to me when I left. I hope you have the opportunity to find such a network of support for yourself.

Everything you said in your letter resonates in my heart. You are right on target. I think your letter was a wonderful gift to your family, friends and parishioners. If nothing else, it got people thinking and talking. You opened the gates to a discussion. The pope may say that we can't talk about it. But the reality is that not only can we talk about it, we must talk about it. And the people want to talk and are talking! You gave a gift of open communication – something very dangerous for the institution that too often realizes it is a dying institution. The church isn't dying, mind you. The people of God are very much alive. There's much right with the church. Thanks for making your decision public in that way. You are to be commended.

Your fatigue and frustration levels seem to have reached the breaking point. When I was trying to figure what to do, I spent a lot of time talking with other priests about how they were handling the same issues. There are a large number of canonical priests in western Washington who have also reached that point and have elected to remain. Some live with the pain. Far too many have turned to dysfunctional coping mechanisms, with unfortunate consequences perpetrated on the people.

I still live with huge doses of frustration and disappointment at the institution. I honestly believe that the church, by being an institution in our day and age, has effectively driven the creative and energizing Spirit to the edges. In some ways the institution has left the church. Maybe the Spirit has been forced to the edges where the people are. So many people I talk to have just given up on the hierarchy coming to grips with reality. That makes me angry, but I know that I'm there too.

From what I glean, you are leaving for a time of discernment. In hindsight I did my discernment and then left. Both are the right way. I think I spent about 18 months in Lent. Advent and Christmas came and went. It was Lent all along in my heart. Maybe you will elect to focus on another liturgical season that best expresses your journey. Please never forget that you are a priest forever, no matter what you choose to do and no matter what others may tell you. Cherish and celebrate your priesthood.

The next few months/years will probably not be easy for you. But I can assure you that this time will be absolutely brimming with the presence of the Spirit. God the Mother and God the Father will open doors that you never imagined possible. But at the same time you will sense doors closing that you had hoped would be new opportunities. All the while, know that God is smiling with you, crying with you, and laughing with you. Our God who is Love is as much with you now as on that morning when you were lying prostrate on the floor and the litany of the saints was being sung.

You are as holy and beloved, as called and chosen now as then. That is the mystery of the Spirit in our lives.

Please know that Marcia and I would welcome the chance to share our home with you if you would like to get away to the soggy Northwest. Call anytime. Stay anytime. If you have the chance to plug into the local Corpus or CITI networks, it may be a helpful source of wisdom for you as it was for me. And know that no matter what path opens for you to follow, I always treasure your kindness and your friendship. Call or write when you have a chance. Blessings….and many prayers.

Peace, my friend,
Bob & Marcia

From the Vatican
January 14, 2002

Dear Father Harpe,

The Holy Father has asked me to reply to the letter which you sent to him on December 19 last, indicating the concerns which led you to take a leave of absence from your priestly duties. It is his hope that your time of discernment will strengthen your faith and strengthen your trust in God's providential guidance both of your own life and of the life of the Church.

His Holiness will remember you in his prayers. Invoking upon you the Holy Spirit's gifts of wisdom, joy and peace he imparts his Apostolic Blessing.

With the assurance of my own prayers, I am

Sincerely yours in Christ,
Monsignor Pedro Lopez Quintana Assessor

January 14, 2002

Dear Dave,

Thank you for your letter updating me on your discernment process. My prayers and support in this important time in your life.

This is an important time for you and your ministerial life. I have always appreciated your commitment to the priesthood and the future of the church. You are right in your letter, the issue has ramifications for the entire church.

My prayer is that the healing power of God be with you in this time. I applaud you taking the time to renew your emotional and spiritual energy in order to make a healthy decision. We expect so much of our priests with little or no support for the journey.

This is extremely challenging times for priestly ministry. We have important questions to ask and yet lack the resolve to reflect in light of our tradition. What saddens me is people like you become the sufferers of unresolved issues. I am sure the past six weeks has not been easy, and so be assured of my prayers and support.

If there is anything I can do, feel free to call or contact me.

Peace and prayers,
Mark

THEY STILL DON'T GET IT
How can a church that judges so many
faithful cover up its own offenses?
by Andrew Sullivan

The Roman Catholic Church holds as an important doctrine that sexuality, though a gift from God, is fraught with moral danger. There is also only one legitimate form of sexual expression – a married heterosexual relationship, always open to the possibility of procreation. So the nature of sex is inescapably bound up with the creation of new life, and any attempt to get around that nature, any variation upon it, is a violation of what God intends. The church therefore condemns masturbation for the same reason that it condemns homosexual sex and contraception. And Catholic doctrine also bars divorce, because it violates the integrity of marriage, which, by definition, can happen only once in a lifetime.

This position has integrity, even though it can seem at times cruel and alien to much of human experience. Most of us know that the force of sexuality, perhaps the most powerful in human life, regularly breaks through such narrow boundaries. But the church insists that its prohibitions are not intended to isolate or wound the divorced, or homosexuals or teenagers in love. Its doctrinal rigidity is maintained out of compassion and besides, the church has no other choice than to uphold the truth, however painful for those caught in humanity's crooked timber.

I suppose it is partly this context that makes the pedophile-priest scandal in Boston so offensive to so many. In the past month, it has been revealed that more than 70 priests in the archdiocese of Boston – out of a total of less than 700 – have been accused of the sexual abuse of children. That's 1 in 10. Worse, when evidence of these crimes has come to light, the church hierarchy has done everything in its power to hush it up, pay secret damages to the victims and, in many cases, do nothing but reassign pedophile priests to other parishes, where they can commit abuse again. In one of the worst cases, that of the Rev. John Geoghan, the church hierarchy had responded to clear evidence of his depravity by moving the now defrocked priest around for almost two decades – as he continued his pattern of molestation of minors. Last week he was sentenced to nine to 10 years for indecent assault and battery.

How can a church that preaches the impermissibility of so many forms of consensual, adult sex simultaneously tolerate, ignore or cover up the sexual abuse of children by its own priests? Pedophilia is not a failing; it is not some imperfect but victimless expression of sexuality among consenting adults. It is a crime. To my mind, the violation of a child's innocence, the betrayal of a priestly trust, the rape of a minor's very body provide about as good a definition of evil as one can find. Yet a church that regularly condemns and judges so many of its congregants for comparatively minor sexual variations of the married heterosexual norm permits and covers up far worse offenses among its own.

And they still don't get it. Yes, Cardinal Law has formally apologized. But his reckoning was carefully parsed. "In retrospect," Law said in his formal statement, the "response of the archdiocese to the grave evil … was flawed and inadequate." In retrospect? By what conceivable moral argument could ignoring child abuse be deemed at any time acceptable? "In retrospect," he also said, he had put children in danger, "albeit unintentionally." How can a church demand moral responsibility of its members if its leaders cannot do so when unmitigated evil is standing right in front of them?

This is not a liberal or conservative issue. Sure, liberal Catholics see the scandal as another indicator of the sexual dysfunction at the heart of the church. And they have a point. Celibacy is an onerous burden that can easily distort a person's psyche. Moreover, many sexually conflicted men gravitate to the priesthood precisely because it promises to put a straitjacket on their compulsions and confusions. Alas, that straitjacket can often come undone. The absence of women in the higher reaches of the church further distorts the atmosphere; and the presence of large numbers of gay priests – forced to preach against their very identity and fight against their own need for love – only intensifies the psychological pressure of the priesthood. But conservatives are just as outraged. The abuse of children rightly provokes horror among traditional Catholics, and they have been admirably reluctant to close ranks behind the corrupted hierarchy. Besides, the most devout and trusting have often been the most victimized. "After he molested me, he would bless me," a former altar boy, abused in the Los Angeles diocese, recently told the Los Angeles Times. "It's very confusing. I was in the center of my mother's life – the church – and she thought I was doing constructive things by being with the priest. After we did these things, he'd put his hand on my head and make the sign of the cross."

This isn't a failing of the church; it's an attack upon its integrity – by its own clergy. Until this evil is rooted out – and until the culpable bishops and cardinals who tolerated it resign – it will surely be hard for American Catholics to trust or love their church again.

6 March 2002

TIIME MAGAZINE LETTERS
TIME & LIFE BUILDING
ROCKEFELLER CENTER
NEW YORK, NY 10020

Dear friends at TIME,

I want to thank Andrew Sullivan for his viewpoint piece, THEY STILL DON'T GET IT, which appeared in the magazine's March 4th issue. Well said! But beyond this horrible scandal I cannot count the times that I have heard or spoken the same or similar words, THEY STILL DON'T GET IT, with regard to the Catholic Church's hierarchy, which, for the most part, is out of touch, out of time and out of gas. For example:

- Command and obey is the posture of the hierarchy. Such an approach might have worked in times past, but not in today's church or today's world. They still don't get it.
- Episcopal ministry has become office management, with Rome directing everything. They still don't get it.
- As regards dialogue within the church, in which the people of God are part of the conversation. Dialogue? You've got to be joking! They still don't get it.
- What about the shrinking number of priests and the stretching of the current supply to the point of ineffectiveness, burnout, or even breakdown? Never mind that the people of God have the right to expect/demand priests who are mentally, physically, emotionally and spiritually healthy. They still don't get it.
- Unnoticed and/or uncared for are the large numbers of persons who leave Catholicism for another denomination, or leave only to have little or no connection to Christianity in general. They still don't' get it.

- And of course, there is the ongoing and desperate struggle on the part of the hierarchy to simply maintain a crumbling structure. A most fitting analogy would be that of the Titanic, having just struck the iceberg. They still don't get it.

On a personal note, I am a priest who is currently on leave of absence from the diocese of Grand Rapids, Michigan. All of these issues have led me to this period of discernment. There is nothing more life giving than the opportunity to minister to God's people AND to be ministered to by them. But the experience is fast becoming life taking within this structure, the leadership of which is concerned only with control and maintenance, rather than freeing a people to live the life of faith. For the most part, I believe the church's leadership body finds itself standing at the threshold of spiritual bankruptcy. Why? Because they still don't get it.

Rev. David L. Harpe

2 April 2002

Dear Bishop,

Greetings in Christ. Back in February I wrote to you about my interest in pursuing what I believe to be a real opportunity for many others and myself. This was regarding the position of chaplain/campus minister at Muskegon Catholic Central. A copy of the same letter was also submitted to the priest appointment committee, in care of Ernie Schneider. I later received a copy of a Xerox letter to Mr. Tom Powers, in which our proposal (that of Tom and myself) was denied because the priest appointment committee felt that "... every available priest is needed in our parishes."

Quite frankly, that response left me with a feeling of suspicion. A phone conversation with Charlie Brown revealed that our proposal was presented near the end of a very long meeting of the priest appointment committee, given little more than a passing glance and then dismissed. I have since had conversation with Don Weber, who said that he would bring the matter up for discussion at a meeting of vicars, which apparently took place last week.

This morning I received a call from Don, who said the vicars suggest I write to you and present a formal proposal for your consideration. They further suggested a personal meeting and/or conversation between you and myself, to which I am most open.

Once again, I am writing with a sense of urgency on the part of myself as well as Tom Powers. Please be advised, however, that Tom will be on spring break during this first week of April. And while writing for the purpose of dialogue, perhaps more than anything else I am asking that we dare to take a risk with this opportunity, one that is filled with promise and potential.

In the midst of the horrible scandal that is unfolding before us, Bishop, I pray with the Evangelist: "The Light shines in the darkness, and the darkness cannot overcome it." John 1:5

Easter peace and joy…always,
Rev. David L. Harpe

Letters to the Editor
National Catholic Reporter
115 East Armour Boulevard
Kansas City, MO 64111

11 April 2002

The church, in all its members, is outraged by the sex abuse scandal that is unfolding for all the world to see. Adding insult to injury, however, is the very leadership of our church.

The bishops have yet to figure out what the rest of us already know, that truth is the only way for the church to move beyond this dark moment. I believe the imagery of St. John of the Cross is most appropriate: We cannot back away from the flame. Nor can we simply walk around it. We must go into the flame, an image of Christ. Only then will we emerge healed, and something of a new creation.

It is reported that the bishops will surely be discussing the current scandal during their June meeting. That might prove helpful, but why wait until June? The bishops should have done everything in their power to have met back in January. They missed a perfect opportunity to truly shepherd a wounded people, by walking with them into the flame. Instead, they remain silent, withholding truth, continuing to deny. And through it all, the bishops seem oblivious to the fact that their integrity and credibility have all but disappeared.

As long as the bishops choose to remain in their present mode of leadership (or lack thereof), we will not wake from our nightmare until one or more of the bishops finds himself presiding over a single-occupancy prison cell. And if that is their choice, then so be it.

Rev. David L. Harpe

18 April 2002

Dear Bishop Rose,

Greetings. This is to acknowledge receipt of your letter of April 12th, and to respond to the questions raised therein.

About the proposal made by Tom Powers, both he and myself were wanting, more than anything, an opportunity to be part of the discussion. That's all. Having now heard from you personally, I accept your decision and consider the matter closed to further discussion. I do, however, feel compelled to comment on your concern that my presence in Muskegon could "...easily lead to the re-creation of St. Joseph Parish on the campus of Muskegon Catholic." This, I'm sure, is a valid concern on your part. But I have a much larger vision of church than St. Joe's, and the people of the parish knew it. This was one of the key elements that was kept before the parish during the discernment process which, ultimately, led them to let go of the parish.

About my personal health, let me say that it is not a "deep" depression that I am experiencing. As I said to both yourself and Dr. Tom L., prior to my departure, I believe there is present some degree of depression as well as feelings of burnout, this following a difficult last year at St. Joseph and then moving into an assignment which I described for you as being something of a nightmare, especially with regard to the parish of St. Bernard. I met with Tom on two occasions, which I realize is not a whole lot of time, but enough time for him to confirm what I already knew about myself. Beyond this, you will need to personally request his written assessment of our meetings.

On the physical level, I am currently under the care of a medical physician who believes that I may have been misdiagnosed in the north. Dr. Ted Alexander is of the opinion that I might have a type of pleurisy around the heart. I am currently trying a couple of medications. If, in another week or so, there are no results, then we will proceed with further testing of a more extensive nature.

On the personal side of things, I have found enjoyment in being of some assistance to my youngest brother who is learning to live with rheumatoid arthritis. I have found even more enjoyment in what has become something of an extensive spiritual writing exercise. This is not only healthy, but quite calming and freeing as well. I had also hoped to attend a spirituality institute, in New Mexico, during this time out. However, all I received from Mark Przybysz was a catalog, which indicates that the next such institute is not until July or August. It would appear, then, that this is out of the question, at least for the time being.

About reassigning me to another parish, I do not foresee myself repeating the experience of departing a parish so abruptly. Not to make excuses, but circumstances were not good one year ago. Nor were there any other assignments available at that time. In a sense, this was like a heart attack that comes without warning. It just happens. And if it is not fatal, the key to survival is to make any necessary change(s), read the signs and, hopefully, avoid a subsequent attack. I do not believe that I was right in responding as I did, but neither do I believe that I was wrong. And further, I do not believe, as was said to me by a few of the brothers, that I was being "given the shaft" or "hung out to dry" (their words) when I was assigned to Irons.

About my statement that I would never again pastor two parishes simultaneously: One must appreciate the context in which that statement was made. I am not closed to the possibility. But this does not mean that I am going to jump at the first such opportunity. In a phone conversation with Charlie Brown, just a couple of nights ago, I spoke about some of these issues. In the same conversation I also indicated that I would definitely be interested in filling the vacancy at St. Michael, Remus. And I am informing you that I would also consider being assigned to St. Michael, Brunswick, to include Christ the King, Hesperia.

About the disagreements that I have with church policy, I am certain that I spoke honestly and directly to these when I visited you at your residence. In a broad sense, it is my belief that the larger church will, sooner or later, be forced to engage in dialogue, a dialogue that must be inclusive of the laity. Further, the church's leaders must be open to the possibility of change. And please, let me be the first to state that change for the sake of change is not very smart. On the contrary, it is pretty dumb. Specifically, the one issue with which I have the greatest difficulty is the continued stretching of priests in order to maintain a male, celibate clergy. And with all of its shortcomings, inadequacies and inconsistencies, heightened all the more in light of the present scandal, the church's leaders continue to refuse to dialogue. There's an elephant in the living room, and nobody wants to talk about it. Priests will often complain about such issues over dinner, or during the course of a monthly deanery meeting. To this type of complaining I have responded: "If you are not as willing to be vocal beyond these walls, then perhaps you should shut up." Maybe this is a bit harsh, but I don't particularly wish to exercise the other option: Complain about issues behind closed doors; and criticize the Bishop privately, while kissing his ring publicly.

On a personal note, one of my siblings (an inactive Catholic) recently said to me "Why don't you just leave priesthood?" It's not that simple. If there were no longer a sense that I am called to priestly ministry, that this is where I am to be, the decision would have been made and I would not find myself writing this letter. But I do love priesthood and the privilege of journeying, as a priest, with God's people.

Finally, Bishop, I extend my sincere apology for the inconvenience that my situation has presented for yourself and others. I am grateful for your prayerful support. And I look forward to our next exchange, as well as my return to active duty.

In Christ,
Fr. David Harpe

As stated in the preceding letter, I met twice with Dr. Tom L. prior to my leave of absence. Not a great deal of time, to be sure, but I was pleased to have had the Doctor voice the following:

- That he believes me to both know and articulate myself quite well.
- That he concurred with my belief that "I am the symptom," and that it is our church's leadership that "has the problem."
- That he could think of a handful of priests who should consider doing exactly what I was preparing to do (i.e., take a time-out for the sake of my own well being).

27 November 2002

Dear Pope John Paul,

It was about this time, one year ago, when I wrote a rather direct letter to you informing you that I am nearly fed up with the current state of leadership within the Catholic Church. The response that I received, from some crusty old monsignor, was more or less "The Holy Father is praying for you." That, in my estimation, is something of a non-response.

In the midst of the sinfulness and scandal that has plagued God's people during the past year, I took a six-month leave of absence from priestly duty. No regrets. Returning to a life that I love, it is very sad to see that the hierarchy continues to forge a path of self-destruction. Worse still, they don't seem to care. We are at crisis point with regard to the shrinking number of priests, yet the hierarchy does nothing. We have large numbers of Catholics who have either ceased to be part of the church or have connected with non-Catholic traditions, and the hierarchy does nothing, save for pointing fingers at anyone and everyone but themselves. God's people hunger for spiritual nourishment, and our leadership points them to a pre-Vatican II well which, by the way, has long since dried up. All these issues, and then some, and here is where the magisterium is directing its energy:

- Kneeling is to be the proper posture during the Eucharistic prayer. So much for embracing unity in diversity! Instead, we focus on uniformity, which is NOT unity.
- No more general absolution. Force God's people into private celebration of the sacrament. Beat them into the Kingdom! Do you really think that that will work? Think again!
- Canonizing individuals in large number and with such frivolity that the process has become laughable. For instance, a Catholic web site once reported a rumor that the Holy Father had canonized an empty potato chip bag. At this point in time, I am inclined to believe it.
- Five new mysteries to the rosary. **WOW!** Now that's what I call problem solving.
- Asking priests who are at or beyond retirement age to remain on active duty as long as possible. Hmmm. Provided I don't take an early retirement or leave the priesthood altogether, I will not work one day past the age of sixty-five. Period!

Some years ago it was reported that a person approached Pope John XXIII and asked him "How many people work at the Vatican?" The pope is said to have responded "About half." Some things seem to never change.

In closing, let me say that I would welcome your response to this letter. But if it is just more paltry piety from yet another Vatican dinosaur, then please do not waste your time. Or mine for that matter. Happy holidays!

With all due respect,
Rev. David L. Harpe

5 December 2002

Dear Pope John Paul,

At this time I am finding it increasingly difficult to even address you as "Your Holiness," for the leadership of our church is a far cry from being "holy." At present, the magisterium seems to be the embodiment of all that is wrong and lacking within the church. To make the point, I draw your attention to the enclosed articles, all of which appeared in The Grand Rapids Press on Wednesday, December 4th. Cardinal Law stands at the center of our woes, but he is also the reflection of a larger body of bishops whose leadership is nothing short of reprehensible. And the fact that he and others have not been driven into the desert – never to be seen or heard from again – or, at the very least, driven into retirement is absolutely unconscionable. Shame on you!

This coming Sunday, December 8th, we will again hear the words of Isaiah: "A voice cries out: In the desert prepare the way of the Lord! Make straight in the wasteland a highway for our God! Every valley shall be filled in, every mountain and hill shall be made low; the rugged land shall be made a plain, the rough country, a broad valley. Then the glory of the Lord shall be revealed, and all people shall see it together; for the mouth of the Lord has spoken." (Is 40:3-5)

If Advent calls us to anything, it is to be that voice. So it is that I will raise my voice, for in faith and in good conscience I can do nothing less. And I will encourage God's people to do the same. We must be the voice crying out to the world for that which only the Lord can give, namely, fullness of life.

In Christ, I am
Rev. David L. Harpe

Rev. David L. Harpe

ADVENT / CHRISTMAS – 2002

Dear family and friends,

Greetings and peace in Jesus Christ. May this letter find you well and joyful in the Lord.

Last year, as you may recall, I experienced a rather difficult moment in the journey of faith. Following the painful yet profoundly spiritual experience of the closing of St. Joseph Parish, Muskegon, and then moving into an assignment where there was spiritual hunger on the part of God's people but spiritual neglect on the part of their leaders, I soon found myself feeling both overwhelmed and burned out. It was time, time for a break. If I were to continue as a priest to walk effectively with God's people in the way of faith, then I would first need some time away. And so it was that I embarked on a six-month leave of absence, during which time I engaged in some rather extensive spiritual writing and reflection, out of which there might soon emerge a book. Stay tuned. As it turned out, I could not have chosen a better moment to be away from priestly duties, although there was no escaping the dark cloud of scandal which has hung over God's people for much of the past year.

To you who are reading this, words cannot begin to express my gratitude for your prayerful support, love and encouragement while I was away. By contrast, only two priests initiated contact with me during my hiatus from ministry, calling and/or writing just to see how I was doing and to offer their support. Who were they? Fr. Tom Simons and (retired) Bishop Joseph Breitenbeck. I share this with you not to be mean, but to simply underscore that which the larger church realizes and its leaders continue to fail to grasp, namely, that priesthood, as we know it is pretty unhealthy. Where there has been sin and crime, there has been payoff and cover-up. But where there is genuine need for support and encouragement, the fraternity is sorely lacking. Perhaps we are stretched to the degree that we do not have the time to give adequate support to one another, or we simply do no know how to provide that support. I remain convinced that it is something of both, and then some. Enough about that, at least for now.

Rested, refreshed and spiritually purged, it was time to get back in the saddle. As of July 10th, I assumed a new assignment: Pastor of St. Philip Neri, Reed City, and St. Anne, Paris. Despite the fact that the parish is a casualty of the present scandal, the people of God are, for the most part, weathering the storm quite well. Down-to-earth, warm, welcoming and spiritually solid, they have made it easy for me to feel at home in Reed City and Paris. At this point, all indicators strongly suggest that we are a good fit for one another.

As church, we are now fast approaching the start of a new liturgical year and the beautiful seasons of Advent and Christmastime. With this new beginning, this "most wonderful time of the year" drawing near, where might we look for light in this present moment of darkness? There is, for the Christian, one and only one source to whom we can look, namely, Jesus Christ and His Word. Despite the sin, the crimes, the scandal, the Lord's voice rises above it all in the sacred texts that are forever filled with promise and the hope of a brighter future. This year there is one Advent text which, for me, seems to rise above all others:

> A voice cries out: In the desert prepare the way of the Lord! Make straight in the wasteland a highway for our God! Every valley shall be made low; the rugged land shall be made a plain, the rough country, a broad valley. Then the glory of the Lord shall be revealed, and all people shall see it together; for the mouth of the Lord has spoken. Isaiah 40:3-5

The Christian community has long cited these words from Isaiah as a foretelling of the coming of the Messiah; It was also John the Baptist who claimed the same words when pointing to Jesus as the Christ. And in time, Jesus himself would cite the prophet Isaiah (61:1-2) when he stood up in the synagogue and proclaimed:

> The spirit of the Lord is upon me; therefore he has anointed me. He has sent me to bring glad tidings to the poor, to proclaim liberty to captives, recovery of sight to the blind

and release to prisoners, to announce a year of favor from the Lord. Luke 4:18-19

Now, during this in-between time in which we live, Advent and Christmas remind us that it was in darkness that God chose to come into the world. And ever since that silent first coming of Christ, every Advent and Christmas further reminds us that we are to be heralds of His glorious second coming. In this present moment of darkness, then, it appears that the people of God must rise to their call, their vocation, with a heightened sense of urgency.

As we prepare for Advent, Christmas and the coming new year, I offer this simple prayer: In the company of John the Baptist and the Lord Jesus Christ, may we claim the words of Isaiah as our own. May we be that voice which cries out "Make straight a highway for our God!" May we be the vehicles through whom the Lord continues to bring "glad tidings to the poor, liberty to captives, sight to the blind and release to prisoners." And in our doing so, faithfully, may the glory of the Lord be revealed.

Peace and blessings…always,
David L. Harpe

17 December 2002

Georgie Anne Geyer
Universal Press Syndicate
4520 Main Street
Kansas City, MO 64111

Dear Georgie Anne,

Greetings from Reed City, Michigan. I am writing to say THANK YOU for the column, Law's resignation offers little comfort, which appeared in the The Grand Rapids Press on Monday, December 16th. Along with my thanks, I am enclosing a few other pieces for your reflection.

I have decided that it is time to speak out, for too many of our priests have remained silent. Further, I have decided to stand with God's people and encourage them to begin raising their voice, not asking for dialogue and reform, but demanding it. In and of itself, this shameful and sinful scandal makes this much perfectly clear: The people in the pews are quite healthy in their faith, while the magisterium are the ones who are really quite sick, standing at the threshold of spiritual as well as financial bankruptcy. Worse still, they (the magisterium) do no know this about themselves. Perhaps through the voice of the faithful, including our non-Catholic sisters and brothers, these old men might find themselves reawakened in and by the Spirit. But given their penchant for power OVER people, as opposed to EMPOWERING people, I would agree with you that there are no indicators that significant change is going to happen.

Once again, my sincere thanks for your words. Have a wonderful holiday and a most blessed New Year.

In Christ,
Rev. David L. Harpe

18 December 2002

Dear Pope John Paul,

I am writing this letter, the third in as many weeks, to share a few more thoughts with you.

In the company of Catholics everywhere, I applaud your acceptance of Cardinal Law's resignation. Quite frankly, it coincided nicely with our celebration of Gaudete Sunday. On the downside, however, I am left to ponder the following:

- Why did you fail to insist on the cardinal's immediate resignation several months ago?
- The resignation of Cardinal Law does not go far enough. I hope the idiot finds himself presiding at Liturgy alone and in a jail cell. In the event that he should manage to avoid any prison time, it would be incumbent upon you to force him

into full retirement, demanding that he simply disappear, entering into a private life of prayer and penance. As the old saying goes: "What's good for the goose is good for the gander." To allow him to continue to function in any capacity within our church would be, at the very least, an affront to the people of God.

- This is not just about Bernard Law. This conference of bishops has no credibility. Every one of them who has been incriminated in the scandal of abuse and cover-up should likewise be forced out. Why they remain is beyond comprehension. It also casts suspicion upon your own ability to effectively lead this church. If I may be direct, it is my firm belief – one that is shared by many – that your leadership has already been compromised by this scandal.

Finally, I want you to know that the people of St. Philip Neri and St. Anne enjoyed a wonderful celebration of the sacrament of reconciliation, with general absolution, this past Sunday afternoon. I will look forward to future celebrations of the sacrament. And yes, these will include general absolution. Perhaps the enclosed article (relating to the sacrament) might serve to open a few eyes in Rome, although I'm not banking on it. At least take a look at the picture, which says it all.

Faithfully in the Lord,
Rev. David L. Harpe

Appearing in U.S. CATHOLIC was a wonderful article: **It would be a sin to lose general absolution**, by Fr. William Stenzel. Gracing this piece, as referenced in the preceding letter to Pope John Paul II, was an image that depicted two different doors to the sacrament of reconciliation: One was marked "GROUPS ONLY," and the other "INDIVIDUALS ONLY." Beneath the sign "groups only," the door was open and people were flocking to the sacrament. The other door, "individuals only," remained closed and shrouded in cobwebs. A picture is indeed worth a thousand words!

It now appears the pope has gotten what he wanted. Once again, it will be his way or no way. And our bishops, kowtowing to the old man who was kind enough to make them bishops in the first place, are now promulgating from diocese to diocese that general absolution is no longer to be practiced as a form of celebrating the sacrament (except under very special pastoral circumstances, which even then will require the explicit permission of the diocesan bishop). How unfortunate. How sad. Such a small and narrow theology will not draw people into a deeper life of faith. It **does**, however, have the potential for closing altogether the very door to sacrament. And that is scary.

As I was writing this particular letter to His Holiness, in the midst of Advent, one of my classmates sent an early Christmas greeting. Along with Mark's personal words of encouragement, there was a most appropriate reflection that I choose to include here, for it speaks well to all of us. However, it is hoped that the message therein will reach the eyes, hearts and minds of a church leadership body that seem all but completely closed...

> Could but thy soul, O Man,
> Become a silent night,
> God would be born in Thee
> And set all things aright.
>
> The light of blessedness
> Shines when night is deep.
> To see it, let thy heart
> Have eyes that never sleep
>
> **Angelus Silesius**

Letters – 2003

3 January 2003

Dear Pope John Paul,

I am ushering in the New Year by writing this letter to you, the fourth since November 27th, 2002.

In a December address to the annual Diocesan Education / Catechetical Leadership Institute, Bishop William Friend of Shreveport, LA., said "Effective leadership in the church requires vision." The Bishop went on to make this distinction between transactional leaders and transformational leaders:

Transactional leaders, the bishop said, "are basically managers who react to immediate situations and pressures, strike bargains with allies and adversaries, follow limited and short-run goals, and seek to maintain equilibrium in what they wish to achieve."

Transformational leaders, the bishop continued, "are to serve as moral agents. They elevate and cause people to rise above their narrow interests." Bishop Friend also spoke eloquently of the need for developing and sustaining a compelling vision for ministry, a vision for the future that is "attractive, worthwhile and achievable." About that vision for the future, the bishop had this to say:

"A vision, quite simply, is a realistic, credible, attractive future for your ministry or organization. It is your articulation of a destination toward which your organization should aim, a future that in important ways is better or more desirable than in the present state. **Vision** is a signpost pointing the way for all who need to understand what you and they are about, and where you intend to go together."

For a brief moment I found excitement in the bishop's words. But I was then reminded that he was speaking to diocesan education and catechetical leaders. While the bishop's words are true, they must first be spoken to the whole college of bishops who, for the most part, exemplify transactional and even **dysfunctional** leadership. Lacking any vision for today, they are incapable of offering God's people a "realistic, credible, attractive future."

With each passing day it becomes all the more likely that I will soon be leaving priesthood. As much as I love it and as passionate as I am about it, I cannot remain silent while this outdated, antiquated and crumbling structure of non-leadership continues to insist on beating-up on God's people, rather than leading and shepherding them toward that which God has already envisioned for us, namely, the Kingdom.

Happy New Year!
Rev. David L. Harpe

7 January 2003

Dear Bishop Rose,

This letter has been taking shape over a period of a couple years, long before any ink was put to the paper. With our celebration of Epiphany this past weekend, I was given something of a revelation about myself. The revelation? That it is time to make a most serious decision for myself and for my personal well being. I have decided to leave priestly ministry, effective no later than the end of January 2003.

Confirmation of the rightness of my decision came with this morning's Liturgy and the gospel reading, Mark 6:34-44. With the disciples feeling spent and their resources limited, they find themselves looking into the eyes of a hungry crowd of at least five thousand. And behold, Jesus turns their limited resources into an abundance of God's goodness and grace. Hunger is not only satisfied, but there are leftovers as well. I then shared with my parishioners what I perceived to be a glaring irony within this scripture text: Today, we are looking into the eyes of a people who are hungry for spiritual nourishment, but our priestly resources are limited. And instead of allowing the Lord to turn these limited resources into an abundance of God's grace, our leadership continues to insist on using our resources for the purpose of corralling God's people and keeping them in line for the sake of keeping the rules. These same rules have not only become the religion, but in the minds of many of our leaders it would appear that the rules have become something of a god. Correct me if I am wrong, but I believe we rightly name this **idolatry**.

Contributing to my decision has been the unfolding of the sex abuse scandal and, further, the years of cover-up, payoff and conspiracy on the part of our bishops, with the latest case (in this diocese) providing front-page reading material in today's newspaper.

Other factors have influenced my decision as well, among which are low morale, cynicism and negativity among priests. In the course of a conversation some weeks ago, one of our priests made this remark: "David, I'm just praying that I can hang on for another ten years and retire." How sad. How tragic. Priesthood should be, at the very least, about finding **LIFE**, not simply a matter of "hanging on." I hope that I am never reduced to such an approach as this, whatever the work I might find myself doing in the future.

The bottom line on all of this is that I have been facing a major dilemma for quite some time: Do I stay, or do I leave? To remain within the present structure is to adopt a business approach to the life of faith, an approach that is more concerned with maintenance and minimalism than helping people to claim and grow in their faith. At a personal level, to remain is to risk having my very spirit broken or even killed, all for the sake of perpetuating a leadership structure that is rot with decay. I cannot and will not do this to myself, or God's people. And to leave is to experience the sadness and pain of having to let go of that which I love passionately. Needless to say, I have found myself at a very difficult moment in time. But it is time.

I've had a couple of people ask how I might respond if there is raised the question of my promise of obedience. That's the easy part in all of this. Given the present scandal and a hierarchical structure that appears to be near totally consumed by self-preservation, all at the expense of God's people, my response to any question about obedience will be quite simple: All bets are off, except for obedience to Jesus Christ and His Word, along with a promise to serve the people of God as best I can in imitation of Jesus' own example.

At this point I feel compelled to make an apology of sorts. And so I apologize to the people of God, especially the people of St. Philip Neri and St. Anne. I am sorry for the disruption and inconvenience that my departure will bring to them. To the larger leadership body of the church I make no apology.

I must remind myself to stay rooted in prayer, for prayer has been and must remain a critical part of the journey. And so I wish to close this letter with the very familiar serenity prayer, highlighting that phrase which impels me at this time:

God, grant me the serenity
to accept the things I cannot change,
courage to change the things I can,
and wisdom to know the difference.

In faith, in good conscience, and in Christ,
Fr. David L. Harpe

8 January 2003

Dear Pope John Paul,

In this fifth letter in as many weeks, and quite possibly the last that I will ever write to you, I simply draw your attention to the enclosed letter to Bishop Robert Rose, a letter that has now been shared with my parishioners and a host of other individuals.

There is much more that I would like to discuss with you personally. But because my words will fall on deaf ears as well as a closed mind and heart, which have all but banned the very use of the word "dialogue," I will save my breath. Instead, I will simply look for other avenues by which I might continue to be of service to God's people, even if it means gathering with them **outside** the rules and regulations of the current institutional structure.

I suppose the only thing that remains to be said at this time is **SO BE IT!**

Again, with all due respect,
Fr. David L. Harpe

The preceding letter is the last of a series that were mailed, in rapid succession, to the Holy Father. Dated 11/27/02, 12/5/02, 12/18/02, 1/3/03 and 1/8/03 respectively, all have gone unanswered.

Rev. David L. Harpe

January 10, 2003

Dear Father Harpe,

I am replying to your letter of January 7, 2003. You wrote to inform me that you have decided to leave priestly ministry, effective no later than the end of this month.

I did not know, of course, that you were having any difficulties this year. However, in view of the events in your life and ministry over the last year and a half, and of the feelings that you express toward the present leadership of the Church, I believe you have made the right decision.

I see no reason to delay your departure until the end of the month. When a priest has made up his mind to leave the ministry, it is usually healthier for his people and for him if he leaves as soon as new arrangements can be made for his parish(es). We will be able to make such arrangements beginning with the weekend of January 18-19.

I therefore am relieving you of your assignment as Pastor of St. Philip Parish, Reed City, and St. Anne, Paris, effective on Friday, January 17, 2003. Your diocesan and provincial faculties will cease on January 17 as well.

If you have any questions about the practical details of your departure, please be in contact with Father William Duncan, Vicar General. If you decide to apply for dispensation from the obligations of the priesthood, we have canon lawyers who can provide advice and assistance.

We are grateful for the priestly ministry that you have carried on since your ordination in 1994. You will be in my prayers as you move into other things.

With every best wish, I am
Sincerely in our Lord,
(Most Rev.) Robert J. Rose
Bishop of Grand Rapids

Sunday evening
1/12/03

Dear Fr. Dave,

My heart aches for you and all you are going through. Our church has many, many problems.

I hope God will keep a close eye on His church here on earth, as man has really messed it up big time.

I just want you to know that I care very much for you and wish you only the very best.

I hope you can head to Texas to be near your brothers. Please know that you will be in my prayers always.

Love and prayers,
Connie

January 13, 2003

Dear Bishop Robert J. Rose,

Greetings and God's blessings. I regret that this is the first time I've ever made the time to communicate with you during your tenure as our bishop.

I write now out of desperation. I am so worried to see what is and is not happening in my faith, the Catholic faith. My family and I have only ever known one religion and I know and believe it is the one true faith. It was never questioned and always respected by my Father, who was raised in the Catholic faith, and it was taught and lived by my Mother, who was a convert to the church. She became an example of the Catholic faith at its best and a picture of respect to all while raising eleven children and working.

My parents are no longer of this life on Earth, but their influence remains strong in all of our lives. I think that this is why it is so difficult to voice my discontent with the state of our Catholic faith.

Over the last ten or so years I have seen numerous families leave the church and go to other faiths where they can receive the spiritual nourishment that they need. These were good people, looking for more from their faith than the Catholic faith is able or willing to provide. Most of these families are involved in their new faith and have lots to offer, and they all indicate that they are seeking a more modern and vital faith to call their own. They also speak of the continued crisis of the Catholic church and the church's refusal to evolve into the 20th century, much less the 21st century.

Vocations have been in serious decline for 30 years, and what has been done to address it? We seem to be looking to other poorer countries to import priests to the United States. I am of the belief that the Catholic religion needs to ordain women and just as importantly, allow our priests and female ministers to be married. How better can they relate to God's flock than to live as their people do and to lead by example. I feel that not to make these changes will only continue the slow demise of our faith and weaken our ability to help the less fortunate countries of the world.

This would also go far to reduce the incidents of priest pedophilia and further eliminate the need for bishops and cardinals to cover-up these criminal acts and the destruction caused to innocent lives.

Most recently I have read and been informed by our priest that we can't receive the sacrament of reconciliation as a parish family and longer. Do the people making these decisions have little else to do or are they doing things like this and other changes just to justify their position? Kind of like job security?

I feel it is important and assuring to participate in reconciliation as a parish family, where we recognize that there are none of us without sin or the need for God's forgiveness.

Why are the church's leaders unwilling (or uninterested) in addressing the major problems in the Catholic faith? I have to wonder if it's not due to their age, complacency, peer pressure and fear of the Pope's disdain.

I fear that if the Catholic Church continues to refuse change or modernization, it will handle evolution about as well as the dinosaurs did.

Should we as Catholics consider ourselves on the "Endangered Species List?"

A Faithful Catholic Christian,
Dennis L. Harpe

April – 2003

Dear David,

I hope and pray that this time of transition has helped bring clarity to your thoughts and peace to your heart. I miss you – I hope you are well. May you know the fullness of Easter joy.

Sue Vallie

24 April 2003
Letters to the Editor
Muskegon Chronicle
P.O. Box 59
Muskegon, MI 49443

A few months ago I arrived at the very sad and difficult decision to step away, indefinitely, from priestly ministry. It seems that my decision has generated a great deal of talk within the Catholic community. This is good, even healthy. And whether one agrees with me or not, this much is certain: Houston, we have a problem! And the problem looks like this:

- For nearly forty years we have witnessed a rapid decline in the number of priests. The result is that we now find ourselves in a moment of crisis with regard to the shortage of **priests** (emphasis on "priest," for there is no shortage of vocations to the priesthood. We are simply not ordaining them).
- The Catholic population continues to experience growth in numbers. At the same time, so too does the number of inactive and/or former Catholics continue to grow.
- Under the present leadership of Rome, the church has returned to a pre-Vatican II structure. As such, the role of the bishop

has been reduced to that of CEO, with priests functioning as branch managers or office boys. And the laity? Church leaders may throw them a cookie from time-to-time, but it is quite clear that the role of the laity is to pray, pay and obey. **Period**!

- We have a pope who, without a doubt, has been a towering figure in the larger world. But when it comes to our own in-house issues, papal ministry is increasingly exercised in near-dictatorial fashion. For example: Despite the declining number of priests and the continued call to at least debate the possibility of married clergy and the ordination of women, the Holy Father continues to insist on a male/celibate clergy only. Further, he forbids any discussion of the matter by priests as well as the laity. And worse still, in the 1990's John Paul attempted to move the male/celibate requirement from the category of **doctrine** (understood as "presumed to be true, but open for debate or discussion") and into the category of **dogma** ("that which **must** be believed by Catholics"). But any theologian who is worth his/her weight, even in manure, knows that A) the Pope is not free to do so on a whim, and B) he failed to meet the necessary criteria for doing so.
- And of course, there is the clergy sex abuse scandal, far from over, which has left our leadership in something of a shambles, with bishops enjoying little or no credibility whatsoever.

These are the primary issues/concerns that have led to my stepping away from priestly ministry. Have I the answer to all the problems that are facing the church? Certainly not. But neither does the magisterium, which offers little more than band-aids where there is massive hemorrhaging. And so I speak out in faith and in good conscience, that the whole church – in all its members – will at least acknowledge the elephant that is sitting in our living room. Then, in the company of our one absolute authority, Jesus Christ, we can return to the **real** business of church, which is about a people who are called to feed one another as we journey together toward the fullness of life in God's kingdom.

Fr. David L. Harpe

30 April 2003

Dear family and friends,

Greetings and peace. This is a letter that I have desired to write over the course of the past few months. I have been struggling, however, searching for the right words. But they do not come. Only recently have I realized that there are no "right words." But what I can do, which perhaps is what God is really calling me to, is to simply share what is within my heart. While every Catholic in the diocese of Grand Rapids is aware of my present status, this particular letter provides an opportunity to address everyone whom I know and love, especially the friends and classmates who are scattered around the country and beyond. And so I write....

If there is one thing I feel compelled to speak to at this time, it has to do with my decision to leave active priestly ministry. Immediately following the decision, which was announced on the weekend of January 11/12, there came a great deal of media coverage and publicity. And while I have no complaints about what was aired and/or printed, I have been concerned about what was **not** made public at that time. Thus I am sharing with you, as I did with the people of St. Anne and St. Philip Neri, a copy of the letter that I submitted to Bishop Rose. Hopefully, this will put to rest any guessing or speculation as to exactly what I might have said to the bishop. As you read the letter, I invite you to call to mind the critical issues that are calling out for attention and discussion within our family of the church.

At this time, I remain at peace with my decision to speak out and, ultimately, step away from priestly ministry. I have done so with the knowledge that I am not alone with regard to the issues that our church, especially its leaders, **must** address, sooner or later. As has been the case, large numbers of the laity have attempted, for years, to raise their voice, even if only to engage in dialogue with church leadership. But they (the laity) continue to find themselves ignored or, worse, "forbidden" to speak out. We also have many a good priest, at a diocesan level and nationwide, who often complain among themselves about many of the same issues and concerns. But publicly, they are silent. Why? Rather than speculating here, I would simply suggest that each of us find the courage to press these good men for **their** response to that question.

While faith and personal conviction have yielded inner peace, I am also feeling a tremendous sense of loss and sadness. Daily, I find myself missing and longing for the life of priest, especially those things that are at the heart of priesthood: presiding at the rites, sharing our stories, growing in faith, journeying together and stopping along the way to celebrate the sacred, which is both around and within each one of us. And so, for the moment, I find myself living with conflicting feelings and emotions. I suppose that that's the way it has to be, at least for the moment.

During this wilderness experience it is necessary for me to live rather frugally and in spartan fashion. This too is all right, for the moment, for anyone who really knows me is aware that beyond good company and chocolate, my needs in this world are few. Shortly after finishing this letter, I will again turn my attention to the spiritual writing and telling of a story that was shared by many of us. This exercise, this project must be completed. My own heart tells me this. And so do the hearts of many Catholics, as well as the inactive and/or former Catholics who have consistently said to me: "Keep writing." Pray that I find a publisher.

I'm afraid it is time to bring this letter to a close. I'd like to do so with the hope that each one of you are enjoying good health, happiness and blessings in the Lord. And as I ask for your continued prayerful support, let me share once more the words of the late Karl Rahner, which continue to provide me with needed strength and inspiration:

The priest is not an angel sent from heaven. He is a man chosen from among men, a member of the church, a Christian.

Remaining man and Christian, he begins to speak to you the word of God.

This word is not his own. No, he comes to you because God has told him to proclaim God's word.

Perhaps he has not entirely understood it himself. Perhaps he adulterates it. But he believes, and despite his fears he knows that he must communicate God's word to you.

For must not some one of us say something about God, about eternal life, about the majesty of grace in our sanctified being; must not some one of us speak of sin, the judgment and mercy of God?

So my dear friends, pray for him. Carry him so that he might be able to sustain others by bringing to them the mystery of God's love revealed in Christ Jesus.

"Pray for him." But let us also pray for one another, for the church that we are, the church that we love. May the new life of Easter raise us up to be the people that God is ever calling us to be in Jesus Christ.

ALLELUIA! Always and with love,
Fr. Dave

A greeting card
Spring, 2003

This is what Yahweh asks of you, and only this: To act justly, to love tenderly, and to walk humbly with your God. –**Micah 6:8**
*

Dear David,

Thank you for the clear and strong example you provide of service to God's people. I hope that both of us will find a place to actively work for and with them.

I continue to pray for you as you work on your book. Please pray for me as I grieve the loss of my role in the parish.

Peace,
Sue Vallie

Dear David,

Thank you for your letter and enclosures. Your story is a very painful one, and one being repeated across the country by both priests and laity. We must, as a church, begin to listen to the Spirit.

We are giving you a complimentary membership in Call to Action and appreciate your support and interest.

Blessings – Sincerely,
Margaret McClory

May – 2003

Hi, Dave.

Good to hear from you. You continue to be in my prayers. I know your decision was not easy for you and I sense your anger and disenchantment.

I, too, have some concerns about the current structure of the church and all the "administration" we priests are saddled with. Yes, it is taking its toll on priests and laity alike.

Having said all that, I choose to remain within the structure and advocate for change. I am not nearly as distrustful of American bishops as you appear to be, although there are some serious judgment issues that surround some bishops.

You spoke in your letter about a lot of low morale and cynicism and negativity among priests. Frankly, I see very little of this. In fact, I find priest's morale generally quite high, although we'd all like to get out from under the yoke of "administration." I've made significant advances toward having my Parish Administrator run the day-to-day operations of the parish.

So, in conclusion, I'd say your experience and my experience of ordained priesthood is considerably different. Please keep this in mind in your writing. I do hope you find peace and nourishment for your wounded spirit.

Larry

Upon receiving the above letter, I immediately wrote to Larry for the purpose of making myself clear on a few key points, while at the same time making a request. What follows is the basic content of that letter:

- I was pleased that he (Larry) and so many others could sense my anger. However, my feelings of anger, etc., are directed at the much larger hierarchical structure of the church, not simply the American bishops.
- "Disenchantment" is not what I am experiencing with regard to the church's leadership. The more appropriate adjective is that of "disgust."
- If Larry has not experienced low morale and negativity among priests, then he is very fortunate. However, polls and surveys (of priests) have consistently reflected quite the opposite.
- Finally, I made this request of Larry: "What, specifically, are you doing to 'advocate for change' from within the structure?"

Well over a year has passed since writing these words. To date, I have received no further word from Larry.

May 14, 2003

Dave,

It is good to hear from you. I'm glad you've found a place of rest and also a place to move on from.

God bless you and your writing project! I know you combine passion with extraordinary talent.

Keep in touch, and peace!
Matt (priest/classmate)

May – 2003

Dear Dave,

For lots of reasons, "Thank you":

- For your ministry among us, which I know from people first-hand, was much treasured.
- For the challenge and affirmation you gave me.
- For pursuing fidelity to principle.
- For including me on your "friend & family" list.
- For continuing to pursue your calling from the Lord.

God bless and reward you,
Lou (retired priest)

May 21, 2003

Dave,

So nice to receive your letter. And thank you also for sharing your letter to Bishop Rose. I can only imagine the amount of soul-searching and discernment that you've undertaken during this past year. Isn't it strange that our work in the church can cause so much pain and frustration? And yet, perhaps, that is the message of the Gospel! When I look around and see God's presence in the world, it truly pains me to see our church caught up in so many things that don't mean anything or aren't relevant to God's people.

And then I hear from you and know that there are good people in this world who "get it." Thanks for writing. Stay well.

Love,
Sue Cook

May 26, 2003

Dear Father Dave,

Yes, we are doing fine. It was good to hear from you. We wondered how you were holding up, considering this difficult time for you (as well as the whole Catholic faith community). Our thoughts and prayers are with you.

Ron and I would definitely be interested in participating in a gathering (with or without diocesan blessing), especially with you officiating at the Mass. The Masses that we attend now leave us spiritually hungry, and never confront the growing questions/ anxieties about the present problems in the church.

Good luck on your book. Keep submitting your manuscript and don't get discouraged. It is a message that needs to be told.

Just tell us "when and where" for the gathering, and we will be there!

Ron, Karen & Danny Pavlich

28 May 2003

Dear Bishop Rose,

Greetings. By way of this letter, I wish to share a few thoughts with you. At the same time, I'd like to ask not so much for a general response, but for your personal and pastoral guidance.

Within myself I continue to struggle with what I can best describe as a love-hate relationship with priesthood and the present hierarchical structure of the church. The love has to do with all the possibilities and potential, all that priesthood **can** be. The hate, on the other hand, has more to do with all the time and energy which priests must give to those things that are **not** proper to priesthood, along with a leadership structure (especially as exercised from Rome) which appears to be not even the least bit proactive, but rather only reactive to the problems and issues that are facing our church. It is truly strange that many of our leaders have much good to speak to the larger world, but within our own household of the church there is little being offered. Worse still, there is no room for question and dialogue. And so our leaders coast along with a maintenance approach to ministry, seemingly oblivious to the growing number of inactive and/or former Catholics.

This is the heart of the struggle. But there are other elements present which further complicate this wrestling match that is not so much between myself and God, but rather between myself and church leadership. These elements are:

- The call to ordained ministry, a call which has come from within the depths of myself and is echoed through the community of believers.
- The belief that good always triumphs over evil, which is akin to our fundamental belief that the light of Christ dispels all darkness.
- The incredible honor and privilege of journeying with God's people, as priest, celebrating the sacred which is both around and within each one of us.

Thus the struggle. And so I ask you, Robert, for your help and guidance, along with your prayerful support and encouragement, for I cannot imagine myself ceasing to be and to live as a priest. Thank you.
Prayerfully,
Fr. Dave Harpe

June 2, 2003

Hi David.

Greetings, my friend. Well, you've made the papers – again (c'mon, most of us get only 15 minutes of fame!). Good letter. I've enclosed a copy in case you didn't get one.*

I've waited to respond to your letter, perhaps because it needs some thoughtful pondering. You and I share some of the same frustrations regarding the institutional church, on different levels and from different perspectives; yours as one who longs to be faithful to the call of the priesthood, mine as one who longs to be a faithful lay person. (But of course, this is only because the prospect of being open to the call of priesthood is one that was not planted in my heart and soul during my formative years. Who knows?) By "faithful" I mean faithful to the gospel of Jesus Christ as we understand it.

That understanding, for the most part, has come through my church. For you as well, I imagine.

So, what do we do? Where do we go? How can we still "be church" without the church? As you know, gathering "outside of diocesan rules" is already happening periodically in Muskegon. With the change in diocesan leadership, I'm not certain how much longer this will be happening.

Yes, I need to ponder this and keep praying for guidance. More later. Just did not want another day to go by without at least letting you know I'm holding you in prayer and hoping for guidance from our good and gracious God. Please keep all of us in prayer as well.

I will end as a friend of min often ends his letters: "I remain with you in the gentle grip of His grace" and from another friend who says "Alleluia always and with love,"

Mary Jeanne

*Letter to the Editor, The Muskegon Chronicle, dated 4/24/03

June 11, 2003

Dear David,

I am replying to your letter of May 28, 2003. It came last week, along with a copy of your letter to the Muskegon Chronicle and a copy of your original letter of resignation that you had sent to some in Muskegon recently.

I was thinking of you last Saturday morning as I ordained three men to the priesthood for service in our diocese, just three days after your ninth anniversary. I thought I would send you a copy of my homily, and make some comments on your letter that may be useful at least for clarification. Today's ordination of priests for the diocese will probably be my last. My Coadjutor, Bishop Kevin Britt, was welcomed in February, and will undoubtedly take over as diocesan bishop well before next year's ordinations. We are enjoying working together.

As I read what you have written in your letter to me and in the latest article, I have the impression that you do not realize fully the consequences of what you have done. You have not just "stepped away indefinitely" from priestly ministry. That makes it sound as if the choice to return is yours, a choice that you can make whenever you wish. That is not the case.

You very publicly resigned from priestly ministry, revoked your solemn promise of obedience to me and my successors, and attacked the teaching authority of the Holy Father, the successor of St. Peter and the highest authority in the Church.

My response to all that, as it had to be, was to accept your resignation, relieve you of your appointment, and revoke your faculties. I also offered to assist you in seeking a dispensation from the obligations of holy orders that you freely took on at your ordinations. "Laicization" would be the logical conclusion to what you have done, and I took you at your word, just as I took you at your word at your ordinations.

As far as the Church is concerned, by your decision and my response you are no longer in priestly ministry or approved for priestly ministry. It surprises me that you are still using the title that belongs to priests in active ministry. That is what you resigned from, and presumably took leave of for good. I realize very well that ontologically you will always bear the special configuration to Christ the Priest that ordination gives. That is not the point at issue.

David, it does not surprise me that you miss priestly ministry. There are undoubtedly personal reasons why you would. What I find missing in your analysis, however, is any sense of the hurt and the anger that you have left behind you in the parishes of your last two assignments. When I visited St. Philip's a month or so ago, I spoke at length with some of the parish leaders. They were not at all happy at the use you made of your position as their Pastor to "have three days in the news" as it was put. Nor were they happy at what they perceived as an enormous lack of comprehension of the suffering they had already been through with the publicity surrounding the removal of their previous Pastor. They were aware that it was precisely that that I had asked you to be most sensitive about. They felt used in a way that they deeply resent.

With regard to your various difficulties with the Church, let me offer a few things for you to think over.

I have the impression that you somehow missed the fact that the Lord chose not angels but human beings to be His Church and to govern His Church. I would suggest a couple years of broad reading in the 2000-year history of our Church. (I am very serious about this) It could be both humbling and consoling. Despite us bishops and priests, the Church takes root and grows. Even in our own country this past year, as the secular press has noted, despite all our troubles, we have had large numbers of converts, while the other mainline churches are steadily declining. Here in our diocese we had four large election rites once again. The Lord does not need us to do His work. He allows us to do it, as long as we do not get in the way too much. As one of my seminary rectors used to say: "It's convenient to love the ideal Church. But the real Mystical Body of Christ is the real Church, with all its human weaknesses." There is no other, and never will be.

Yes, there is a growing shortage of priests. But that is hardly new. It has been talked about for at least twenty-five years now, and you must have been well aware of that during your seminary years, if not before. I have to believe that you were also well aware of the Holy Father's position on celibacy in the Latin Church (a law) and his teaching on the ordination of males only (a doctrine). He has expressed that position and that teaching for twenty-five years now, following the teaching of Paul VI and Blessed John XXIII before him. It is hardly new or a sudden whim.

On the practical side of that issue, we have figures from the Christian Reformed Church, the Episcopal Church, and the Lutheran Church that show that they are facing similar shortages, despite having both married clergy and women clergy. The CRC, for example, faces a minimum of 20 percent of their congregations without pastors at any given time. Doing away with celibacy and ordaining women will not solve the shortage. What we need is a renewal of strong Christian family life, from whence vocations naturally flow.

With regard to the sex abuse scandal, our priests by and large have been very slow to criticize the American bishops for their handling of these cases. Most of them understand two things: (1) It is priests who have caused this whole affair in the first place, by abusing young people; (2) It was in trying to protect their priests from losing their priestly ministry, and people from scandal at priestly conduct, that bishops have handled these cases the way they have. It is odd to have priests berate us for being too protective of brother priests.

The call to ordained ministry in our Church is formally expressed by the Bishop on behalf of the Church community. It takes into account the suitability of the candidate for ministry as well as his desire to seek ordination. The Church has the right to set criteria on which to judge the suitability, since the Church community has the right to be protected against incompetent or harmful ministers. We were just reading about that again in the Epistles of St. John during this final part of the Easter season. It is not new.

I have laid things out as I see them, as plainly as I can, not to cause you pain, nor because I am angry with you, but because I want you to understand how difficult it would be for you to be accepted back into active priestly ministry, either in Grand Rapids or elsewhere. You must realize that no bishop in the country will take a chance on a priest with doubtful background at this time in our history. We are all suffering because some have done that in the past.

You are of course in my prayers, David, as I hope I am in yours during this time of transition in my ministry. Write me again, if you wish, after you have reflected on what I have said. If you are ever in the north, I would suggest that it would be good for you to meet Bishop Britt. He will soon be taking over my office. He is a good and compassionate man.

Sincerely in our Lord,

(Most Rev.) Robert J. Rose
Bishop of Grand Rapids

Reflections On The Preceding Letter

Despite my having read the bishop's letter numerous times and having prayerfully reflected on his words, I find the following to be worthy of mention:

- In my letter of May 28th, 2003, I asked the bishop for his personal and pastoral guidance. With each reading of his letter of response, the bishop's words come across as a three-page chastisement. Void of "personal and pastoral guidance," the bishop's response, ironically, seems to lack even the slightest hint of compassion.
- The general tenor of the bishop's letter seems to suggest, once again, that it is me who has a "problem." And once again, I assert that it is the church (the whole body) that has some serious problems and that I, along with a growing number of others, bear the symptoms.
- I found it somewhat insulting that the bishop would suggest "a couple years of broad reading in the 2000-year history of our church." I would be open to such a suggestion, but only under one condition: The majority of the college of bishops, including the Bishop of Rome, must be sitting beside me in the classroom.
- With regard to the sex abuse scandal, the nation's bishops have been accused of participating in the further scandal of cover-up and conspiracy. It would appear that this accusation is given credence when, in his own words, the bishop states: "It was in trying to protect their priests from losing their priestly ministry, and people from scandal at priestly conduct, that bishops have handled these cases the way they have."
- Also, the bishop has written "It is odd to have priests berate us for being too protective of brother priests." Yes, I would agree. But at the same time, when a building is in flames is

it not incumbent upon someone, anyone, to shout **'FIRE!'**, and then summon the appropriate help?

- I found it particularly harsh, and even offensive, that the bishop's words seem to suggest that I am incompetent or harmful. Given the present climate within the church, I will leave it to the reader to determine **who** are the "incompetent or harmful ministers" in all of this.

- Finally, the bishop makes this statement: "...I want you to understand how difficult it would be for you to be accepted back into active priestly ministry, either in Grand Rapids or elsewhere." Difficult? It seems that we have quite a history of welcoming men back into active priestly ministry after having committed far greater offences than that of publicly criticizing the church's leadership. But I am to be blackballed. Hmmm. I find this to be more than simply puzzling. It is truly mind-blowing.

June 23, 2003

Dave,

The Catholic Church and its priests are struggling once again in St. Louis. Oh, to have people of faith lead the church would be a good thing. Integrity would be good too!

May God richly bless you as you serve God "with a new song."

Peace and Love,
Grace

June – 2003

Dear Father Harpe,

I apologize for taking so long to return this letter. My life is so busy these days I wonder what direction I am going at times.

I was very touched and moved by your letters. They express a sincerity of spirit that is rare these days and seems to be rarer in our priests.

On a personal note, I am not sure I told you, but I am a therapist in private practice for the last 9 years. I am a cradle Catholic and currently the head of my small rural church's Stewardship Committee.

When I read your letter I think about all those clients I know that truly seek enlightenment in their lives and wake up one day and say "…this is not enough and I am no longer fulfilled by the choices I made. I need to make a profound change." I congratulate you on your strength and will and faith. Few people have the strength of character to make the very difficult choices that life gives our way. You have done this, although I know it is like walking into an unknown forest and not knowing what one may discover or what peril runs ahead.

I might make several suggestions to you for reading that, if you have not already read, you might find helpful. I suggest first that you read Plato's Allegory of the Cave (if you read it a long time ago, you might read it again). I also suggest the Celestine Prophecy, the 10th Insight, and several or any of the Dali Lhama's books, which I find helpful.

I have done several papers and research on the Catholic community and the Catholic priesthood as part of my Ph.D., and I have developed some interesting yet rebellious insights.

I no longer consider myself a Roman Catholic. I am an American Catholic. I also know that this institution we call church has been bastardized from what Jesus originally intended, and he would not recognize this community as the one he had hoped to pull away from the old religious thinking. He had been educated among the Essenes and as such, had some rather rebellious ideas of his own.

I remind our pastor constantly that this is NOT his church, and that this community belongs to the people. He needs to let us make this faith community our own. Unlike many of my peers who have fallen away from the church for a variety of reasons, I have never stopped going to church and participating in the sacraments. But I have done so because of my own selfish reasons. I like belonging to a thing that is bigger than me; I like standing at the altar to give Eucharist.; I like knowing I can go to any Catholic church in the whole country and know the prayers and the songs and feel at home. Despite the stress of a constantly changing world, this one thing has remained a constant in my life.

My faith in God has not always been as constant, however. God and I have, at times, agreed to disagree on some things and although my relationship with him/her is perhaps very different than most Catholic Christians, I know it has provided me with a strong net to hold me up even in the most difficult of times.

With that said, I have to tell you that I refuse to pray for vocations to the priesthood. I pray that there will be no new male priests and that we gain enlightenment, not vocations. This crisis has occurred because this patriarchal institution no longer serves the needs of its people. It needs to change.

Remember the old story about the man whose home is flooded and he is on the roof praying for God to help him? Well, when he finally drowned he asked God "Why didn't you save me?" God said: "I sent a boat, a helicopter and a raft. What did you expect?" Well in the same way, this church is not paying attention to the message. Married men and women need to be allowed to be priests, and the laity must be respected. Moreover, women and their role in the church must be respected. Without us, there would not be a church, there would be no children baptized, there would be no music, religious education, or altar servers. It is women who have kept this church going for so long, and it is time our role must be respected. If we ever decide to rally together, this church will come crashing down.

Women will not tolerate their children being abused, or this new generation of women will not allow themselves to be marginalized. I am not the only one who thinks this way. There is a growing movement of women who will no longer stand for what is going on in the Catholic Church in the U.S. today.

I hope you will take the time you need to find the answers that I know are within you. I hope that you will emerge from this time knowing what you did was correct for you and that you can do no good for anyone when you are not emotionally, spiritually, physically and psychologically strong. But we need good people on the inside, fighting the good fight. We cannot lose forever all of those who have become frustrated and angry. So be well and find peace, and come back to minister to those of us who understand this struggle. THIS CHURCH will change, and I intend to be part of it. I think we need you too, Father Harpe, to find your place in this rebellion.

Peace,
Christine

8/27/03

Dear Father Dave,

Thank you for sharing in your letter thoughts that so many of us are trying to articulate. This week's scriptures are so clear in Christ's frustration and anger with the Pharisees and Sadducees as shallow hypocrites. When our church, which should be the bulwark and bastion for protection of children, chooses to protect the "old boy" network rather than take a stand against a heinous situation, we deserve the "woe." I hear so many of my protestant friends echo the same sentiment. "Churchianity" has superceded Christianity. I am hopeful that all the American priests who are now standing tall for an option to celibacy will be heard. If they are condemned, I fear all the talk about an American schism could actually happen. My home parish in Ohio has two priests; one married with children, the other single. Where is the logic, justice or theology in this?

I do have faith that our Lord will prevail in this attack of evil. It is so much greater than my ability to comprehend. Yet if the great command is to "Love your neighbor as yourself," Christians / Catholics, no matter what the status of ordained or lay, must live daily lives of servitude to others in need, rather than propagate the organization for its own sake.

The Muskegon Deanery is in great pain. Some of it is natural, due to change. Some stems from the same lack of empathy that closed our beloved St. Joseph, never counting the cost or seeing the priceless value of a church that was a gathering or rich and poor, young and old, all the races. I wonder what did happen to all the "sacred vessels" and their jewels. The roof was far more important.

As for MCC, we will have four seminarians in the last five years, in spite of a world culture of debauchery rather than dignity. I personally am facing the anxiety of where God wants me at this time next year. The Diocese has announced that my position is eliminated. They are moving to one Diocesan Board for policy, and one superintendent. They will have five regional councils or boards with some degree of regional leadership. I doubt that I am asked to serve. I seriously doubt that I want to serve in that capacity.

Cathy and I will indeed pray that Bishop Britt will hear you out. My first encounters with him have been positive. He is a decision-maker in limbo right now. Please know that if you want us to, we would be honored to write or call him on your behalf. I wish we could do more right now, financially, for you. We just sponsored two African American girls to MCC. Their books alone cost over $300.

You are part of our lives and our community. Despair and loneliness run hand in hand. God will send His Spirit to guide us. I am always reminded that all the prophets and all the apostles saw horrible pain in their lives. Accepting being faithful takes real courage.

His peace,
Tom

5 August 2003

Dear Bishop Britt,

Greetings. I find myself in a rather sad and even painful moment in the journey of faith. As I'm sure you are aware, I resigned from priestly ministry in January of this year. I don't believe it is necessary, at this time, for me to expound on all that went into that decision, for I assume that you are privy to all such information. However, I now find myself rethinking the seriousness of my decision, wanting desperately to return to active priestly ministry.

Time and distance have a wonderful way of forcing one to reflect. Strange as this may sound, by becoming something of an outsider who is looking in, I am able to better appreciate the imperfections that are seemingly part and parcel of any structure, including that of church. Further, I am realizing that I (alone) have little or nor control over many of these issues which can change only with time and by the grace of God. Perhaps the greatest pearl of wisdom that is opening before me is that the world and even the church are imperfect. And it is precisely there, amid all the imperfections, that God calls to each one of us to live, breathe and minister, taking great care to make sure that it is the Lord, not David Harpe or anyone else, who is leading and changing us along the way.

While having made recent mistakes in judgment, which may have caused pain and distress for others, I believe that I have offered (as priest) much more that is good. I believe, just as strongly, that I can continue to offer much good (as priest) in service to the church. And so I am writing with the request that I be allowed to return to priestly ministry within the diocese of Grand Rapids. I am writing to you, Bishop Britt, because it is apparent that you will soon be our bishop.

I look forward to hearing from you. And I close this letter with my prayerful best wishes for you, for Bishop Rose and the church of Grand Rapids.

In Christ,
David L. Harpe

5 August 2003

Dear Bishop Rose,

Greetings. I appreciate your letter of response, dated 11 June 2003. During the past several weeks I have read your letter many times. Having allowed sufficient time to absorb and digest your words, I feel there is more that I must say to you.

For the most part, your letter seemed very harsh. I have no problem with that. However, it would seem that you took my words and actions as a personal attack on yourself. This has never been my intent. Once again, when things erupted for myself, I did my best to make it very clear that I was not attacking you, but rather a systemic problem within our church. If, however, any of my words were a direct assault on you as a person, then I am deeply sorry, and I am in need of your forgiveness.

As regards your comment about the parish leaders at St. Philip and their "having felt used;" those were the words of one person whose name I will withhold. The same person, however, has been very critical of you. And on more than one occasion, I cautioned this individual against personal attacks when he is really speaking about systemic problems. Also about parish leadership, there was little at both St. Philip and St. Anne: No pastoral advisory council, and a shared liturgy commission which seemed to gather only for the purpose of asking: "What does Father want?" Despite his being well loved, there was much that Fr. Dan did not tend to within the parish. Everything spoken to in this paragraph was addressed to the parish prior to my departure. Difficult? Yes. But we had heard each other on all of this.

Unless I am misreading your letter, it struck me as particularly odd and especially harsh that you would refer to me, in the midst of our scandal, as being "incompetent or harmful." All things considered, and certainly not to minimize my actions, I simply added my voice to those of others, some of which are the voices of priests. I know that I spoke similar words to you, one year ago, at which time I said: "Far too many of the brothers criticize and malign the bishop privately, while kissing his ring publicly." That is not the kind of relationship I would wish to have with anyone.

I would like you to know that I have written to Bishop Britt, as per your suggestion. Where it will lead, only God knows. In closing, however, I want you to be assured that you have remained in my thoughts and prayers. I hope that I may count on yours.

In Christ,
David L. Harpe

3 September 2003

LETTERS
NATIONAL CATHOLIC REPORTER
P.O. BOX 419281
KANSAS CITY, MO 64141

I am writing, with great joy, to applaud the 163 priests (from Milwaukee) who recently signed a letter to the U.S. Conference of Catholic Bishops, asking that the discipline / requirement of celibacy be lifted. This is good news, to which there is only this most fitting response: **ALLELUIA!**

This is a beginning, a first step. But I am compelled to offer my brother priests a word of caution (in the event that they have not already given this their thoughtful consideration): Prepare to encounter a wall of resistance in the form of a structural system which does not lend itself to debate, nor respond kindly to those who would dare to question and/or challenge that system. Case in point: Five months after resigning (in protest) from priestly ministry, and having publicly spoken to any number of issues which the church **must** address, my own bishop informed me that I "...would not be allowed to return to active priestly ministry in (his) diocese, or anywhere else for that matter" because I am "incompetent and harmful for the church." Hmmm.

Thank you, priests of Milwaukee, for having the courage to speak out. Suddenly, I do not feel so alone. I can only hope that your example will spread from diocese to diocese, giving other priests the push they need to publicly speak out in faith, truth and love. And with the wall now crumbling, I hope our bishops will find the same courage that will allow them to enter into faithful dialogue, on any number of issues, with the whole people of God. I'm afraid they cannot afford to do anything less.

Fr. David L. Harpe
(priest in self-imposed exile)

So much for courage and dialogue on the part of our bishops! Apparently, the possibility of even discussing optional celibacy was not on the table at their recent fall conference (November, 2003).

In The Dallas Morning News' article, **Bishops won't discuss celibacy**, by staff writer Susan Hogan-Albach (November 12, 2003), there emerges what can best be described as a tale of two bishops, one who represents the vast majority of the college, while the other provides an ever-so-slight glimmer of hope. In the article, Hogan-Albach writes the following: "Most bishops aren't open to talking about allowing priests to marry because Pope John Paul II flatly opposes it, said Bishop Donald Kettler of Fairbanks, Alaska. 'We've got to do what the boss wants,' he said. But Bishop Michael Pfeifer of San Angelo said the 'boss' isn't the pope, but God. A discussion about celibacy could be beneficial, he said. 'We don't lose anything by listening.'"

I doubt that the reader will have any difficulty making the distinction between these two bishops. And so it goes. The gulf only continues to widen between a tiny little group of wanna-be businessmen and the considerably larger group of the faithful (the laity) who are also church.

September 7, 2003

Dear Father Dave,

Jim and I were so happy to receive your letter. We're sorry that things are kind of hard for you, now. Please accept the enclosed gift as a token of our loving care for you as a member of our church community.

Jim and I went to Christian Community Center before St. Joe's and only left there because we felt we should look for an actual church community. Since the closing of St. Joe's, we have drifted back here. Now we have made a few friends, and we enjoy the services very much. I believe Father LaGoe wanted to retire this year, but some of the people, I think from St. Joe's, convinced him to continue for a while. There seems to be a small staff at the center running the meal programs, after-school programs and charitable programs. Father LaGoe also goes to the prison each week, and I believe that ministry means a lot to him.

We're excited to hear that you may return to the priesthood. I know, however, that God will decide where he has need of you. But still, we would love to have you return to Muskegon.

You entered our lives when we were at a time of questioning faith, and you helped us reaffirm our belief in God and His church. We are so grateful for priests like you and Father LaGoe, who, through their strong personal faith, inspire others to believe.

We pray for you and wish you the best. We are sure that God is working something out in you and the He will lead you to where He has use of you. Hopefully, it will be here.

With love,
Jim and Anne

September 19, 2003

Dear David,

I am replying to your note of September 7, 2003. I apologize for not responding to your earlier letter. However, I was away for Spanish studies and away again for prescheduled meetings (from prior to my arrival in Grand Rapids). A good deal of the past month I have been away. This coming week is the annual Priests' Convocation as well.

I am not completely familiar with your situation as much seems to have occurred around the time of my arrival in the diocese, when my attention was on many new things coming my way. Bishop Rose, of course as bishop of the diocese, was in contact with you. Though I know some of the situation, I do not yet know all the details.

Upon my return from the Convocation I will be able to look more carefully into the situation and respond to you as soon as possible.

Thank you for your patience.

With all best wishes, I remain

Sincerely yours in Christ,

Most Reverend Kevin M. Britt
Coadjutor Bishop of Grand Rapids

15 October 2003

Dear Bishop Rose,

Greetings. Well over a year ago, in the winter of 2002, I presented you with a written proposal and request for allowing me to work as full-time chaplain at Muskegon Catholic Central High School, while at the same time serving as a weekend supply priest, primarily within the Muskegon deanery. Both the proposal and the request, supported by a significant number of persons and priests of the deanery, were rejected by you for the following reasons: 1) Concern, on your part, that I would "reestablish the parish of St. Joseph on the grounds of Catholic Central," and 2) as stated by yourself, "every available priest is needed within our parishes." Although I disagreed with you, in writing, with regard to your concern that I would "reestablish the parish of St. Joseph," I accepted your decision. It was time to move on.

Some weeks ago, however, it came to my attention that Phil Sliwinski is now enjoying that which I was denied, that he is serving as chaplain at Muskegon Catholic Central, while at the same time making himself available as a weekend supply priest in Muskegon and/or elsewhere. This leaves me rather perplexed. As such, I cannot help but wonder as to why. What has changed? What has happened to "we need every available priest in our parishes?" And so I am raising these questions, Bishop. Please know that I am doing so not for the sake of confrontation, but for clarity and understanding.

I would imagine that this is a busy time for you, as the diocese finds itself in a moment of transition. Nonetheless, Bishop, I would appreciate your response. In the meantime, I do continue to remember you in prayer. I hope that I may count on yours as well.

Peace...always,
David L. Harpe

21 October 2003

Dear Don,

Greetings. I hope this letter finds its way to you, for I do not know where you are residing since your retirement from ministry.

The journey and the struggle continue for me. For nine months I have had the vantage point of being something of the outsider who is looking in, which has proven helpful in putting things in proper perspective. I've come to a greater appreciation of that which I have always believed with regard to priestly ministry, namely, that there is much more good than bad and, without a doubt, more light than darkness. And looking at the bigger picture, the whole experience of the past nine years, I remain firm in the belief that I am called to priestly ministry. Further, I now want to return to active priestly ministry within the diocese of Grand Rapids.

So, what has changed? I would have to say that it is I who have not so much changed but rather, grown. I have grown to realize that the very things that led me to walk away would still remain if I were to return to active ministry. But I have also come to know that while the priest – any priest – is committed to speaking the truth, he must take great care in allowing change and transformation to occur in God's time and on God's terms.

All of this has been communicated to Kevin Britt, who I understand is now the Bishop of Grand Rapids. I am writing to you, Don, because I have been told that you remain the Vicar for Priests. As such, I am asking for your support as I seek to return to active ministry within the diocese. Please know that you are welcome to share this letter with Bishop Britt. Should you have any questions, you can reach me at the address and/or phone number below.

I look forward to hearing from you, Don. In the meantime, I hope that you are enjoying good health and a well-earned retirement.

Peace…always,
David L. Harpe

To date, there has been no response to the preceding letter.

5 November 2003

Dear Bishop Rose,

Greetings in Christ. I am writing to add my voice in thanks to God for your service to the church and to wish you the very best as you begin yet another chapter in your life.

I continue to peck away at a writing project that was begun well over a year ago. All the indicators seem to suggest that now is the right time for me to bring the project to completion and get it into the hands of a publisher. At this particular point in my writing, I am in the process of documenting something of my personal spiritual journey during the past couple of years. Toward that end, I am also writing to extend to you the courtesy of informing you that I wish to include in my writing three or four letters which you had written to me over the course of this journey.

The journey continues, always. And so does the process of ongoing discernment as to where it might be that the Lord is calling any one of us at any given moment along the way. As both of us continue to prayerfully listen, reflect and discern, may we support one another well, that we may follow the Lord faithfully. May the Lord continue to bless you with good health, happiness and peace for many years to come.

Sincerely yours,
David L. Harpe

Personal Thoughts / Observations

In the preceding chronology of letters, especially those pertaining to this writer's personal journey, attention is drawn to the fact that little is heard from the voices of priests. Consider the following:

- At the time of my resignation, in January of 2003, a number of priests from the diocese called to offer encouragement and to say "we're with you," or "we support you." Since then, however, they have remained near totally silent.
- Beyond Grand Rapids, and also in the course of the past year and a half, I have heard from only a couple of my seminary classmates (which numbered thirty plus, from twenty-two dioceses).
- At a very broad level, priests tend to remain silent. For example: Just prior to Cardinal Bernard Law's resignation in disgrace, in December of 2002, nearly sixty priests of the Boston archdiocese had signed a letter recommending that the cardinal step down. It must have taken a great deal of courage to sign their names to such a letter. But what about the other few hundred priests of the archdiocese? Where were they? Given the circumstances that led to Bernard Law's departure, how could any priest justify having **not** signed his name to that letter of petition?

  During the past two years, laypersons have repeatedly stated to me: "Father, you are about the only priest I have ever heard speak directly to the issues that are facing our church." And in the same breath, most will ask this question: "Why are so many priests silent?" I am not the person (or persons) whom it will take to adequately respond to that question, which is incredibly complex. However, I offer the following as pieces of a very large puzzle:

- Some priests have found that they enjoy the "business approach" to church. As such, they are not about to say or do anything that might anger the boss or, worse, jeopardize their future in the "company."

- Other priests recognize, as do most persons, that we have some serious problems within our ranks. But given their age and experience, they choose to not "make waves." Instead, they opt to ride the clock to retirement, though continuing to grumble from behind closed doors about all that is "wrong with the church." With the internal fire of faith and excitement nearly extinguished, they focus their energy on that which is immediately around them.

- And then there is FEAR. Talk about opening a can of worms! Quite frankly, this is the fear of being found out, fear that the darker side of one's humanity might suddenly be exposed. Within priesthood we see something of that darkness manifested in very common forms:
  - o Addiction (e.g., gambling, drugs, alcohol, etc.).
  - o Embezzlement.
  - o Love affairs (heterosexual or homosexual).
  - o Personal lawsuits against priests (relative to the above activities or behaviors).

Incidentally, in many instances bishops have had a history of addressing these issues/problems in much the same way as they have previously dealt with priest pedophiles, that is, by simply moving the priest(s) to another parish.

Having shared these thoughts, I do not wish to leave the reader with the impression that this is a general representation of priesthood. True, far too many of them remain silent when their voices need to be heard. But I strongly believe that the majority of our priests are good and dedicated persons of faith who, day in and day out, continue to give generously of themselves in service to the Lord and His people. And despite the present moment of darkness within our church, many priests continue to shine brightly. They are indeed worthy of our love, support and admiration.

Near the beginning of this chapter of our story, I made mention of the clergy sex abuse scandal and the further scandal of cover-up and conspiracy on the part of our bishops, relative to which I went on to state that Rome is "very likely to have been involved all along." All things considered, one needs only to ponder this question: How could Rome (specifically, the pope and the Roman curia) have **not** known? Still, as a way of qualifying the above statement while at the same time inviting and encouraging the reader to further reflect on the matter, I will offer a few comments.

Church law requires that each diocesan bishop, scattered around the world, travel to Rome every fifth year. In the course of his ad limina visit, as they are called (short for ad limina apostolorum, or "to the threshold of the apostles"), the diocesan bishop is "bound to present a report to the Supreme Pontiff every five years concerning the state of the diocese committed to him..." (canon 399). With this in mind, there is one obvious question that could (should?) be asked of every bishop whose diocese has been rocked by the clergy sex abuse scandal: Tell me, Bishop (name), during any of your ad limina visits, did you share our dirty secret with the Pope?

If bishops were to respond "no" to this question, then they are left with even more egg on their faces for failing to bring themselves to speak honestly and truthfully to the Holy Father. But if the bishops (even just one of them) were to respond to the same question in the affirmative, that is with a "yes," then it is pretty safe to say that word has consistently come from the very top for bishops to "keep the lid on this mess."

As we move through the nightmare, which will take years, may our hearts, our thoughts and our prayers be united first around all who have been victimized at the hands of priests. For the entire church, the body of Christ, may we find healing and newness of life. And for the leaders of our church, especially the bishops? At the very least, let us hope and pray that the magisterium will again learn this simple yet profound lesson, which is as old as Adam and Eve: Truth has the damndest way of emerging, sooner or later.

A July Gathering

As previously stated on a number of occasions throughout this story, many Catholics are finding themselves pushed to the edge in their search for spiritual nourishment. Nationwide, pockets of spiritually hungry Catholics are looking to some rather unconventional ways of gathering as church, often outside diocesan rules and regulations. Even in Muskegon, Michigan, such occurrences are known to be happening. Again, this is the present reality: God's people are hungry, and the institution is failing them...miserably.

In response to a people who, like myself, are crying out for spiritual nourishment as well as reform and renewal within this church, of which we are a part, the following letter went out during the summer of 2003.

22 July 2003

Dear _____,

Greetings. I may have spoken directly to you or I may have left a message on your answering machine within the past couple of days with regard to the possible gathering of myself with a number of Catholics who, along with myself, share many concerns about our church in its present state. This note is to inform you that I will gladly meet with any interested parties on TUESDAY, JULY 29th, AT 7:00 P.M. AT MUSKEGON'S HERITAGE LANDING.

I realize that this meeting is announced with very short notice. This I could not help. Nonetheless, if you wish to share this information with others, please feel free to do so. We must be clear, however, that this particular gathering is NOT about decision-making. It IS, at this time, nothing more than an occasion for dialogue. And to assist in that dialogue, I offer an agenda of sorts at the conclusion of this note.

I hope these words find each of you well in the Lord. And I look forward to seeing you on July 29th.
God's peace and blessings... always,
Fr. Dave

AGENDA

- Welcome
- Prayer
- Personal Reflection (Fr. Dave)
- Sharing of our stories, our journey of faith (preferably in small groups), keeping in mind the following:
  o **Who** are we, as church?
  o **What** are we to be about as church?
  o Am I (are we) being fed and nourished, spiritually?
  o What is our Theology about ourselves? About church? About God?
- What would I (we) like to see happen as a result of this gathering?
- Other things to consider:
  o Change
  o Letting go
  o Practical concerns
  o Am I ready to commit?
- Miscellaneous
- Is there to be a next step?
- Closing prayer.

On rather short notice, some twenty persons – of varying backgrounds and lived experiences – gathered at Muskegon's Heritage Landing on July 29th, 2003. Though small in number, our gathering was in many ways huge. For the most part, it was an overwhelming confirmation of that which, for a growing number of Catholics, is the present reality within the church:

- We've got many problems.
- We hunger, but we are finding no food.
- We have no voice.
- We are being pushed to the edge.
- We are open to looking elsewhere for nourishment.
- Etc., etc., etc.

Perhaps the most telling aspect of this July gathering was in the fact that a few individuals expressed a readiness to make a split of sorts. The more accurate term would be "schism." These individuals, speaking honestly and directly from their own experience of pain and discontent, serve to underscore the present reality, calling it what it is. A formal schism has not yet been declared within Catholicism. But when growing numbers of persons are leaving the household of the church, "splitting" if you will, what else is one to call it? There IS a split occurring within our church.

Others, including myself, found that we have not yet been pushed so close to the edge that we are ready or willing to make what would be a most painful, difficult and conscious decision to leave Catholicism. After all, the church has been – and it remains – our home. And as the human heart knows, all too well, seldom does one take the experience of "leaving home" lightly. This too is very telling.

So, was there to be a next step? Yes. The general consensus was that we should proceed in this way:

- Remain solid in our faith and convictions.
- Stay rooted in scripture and prayer.
- Continue the dialogue, inviting others to join the discussion.
- Find and raise our voices.
- Hope for, expect and even demand (of the institution) reform and renewal within the church.

Personal Notes on this Priest's Journey

The last three letters that I have written to Bishop Robert Rose (dated 8/5/03, 10/15/03 and 11/5/03 respectively) remain unanswered. In mid-October, Bishop Rose retired and was immediately succeeded by his coadjutor, Bishop Kevin M. Britt. On Wednesday, December 10th, 2003, I met with the new bishop for the purpose of discussing the possibility of my returning to active priestly ministry within the diocese of Grand Rapids. I am pleased that Bishop Britt was open to meeting with me and listening to my personal journey and struggle of the past few years. Perhaps it goes without saying, but it was necessary for me to also do a great deal of listening on December 10th. There is nothing wrong with that, for listening is probably the most critical component of any dialogue. Now I must patiently await the bishop's decision, which will come in a matter of weeks. How appropriate that I should be writing these words during the third week of Advent, that season of the church year that is ever calling and inviting each one of us to prepare for and **wait** for the Lord.

With Christmas little more than one week away, and many of us crying out for something of a new birth in the life of faith and perhaps especially within the life of our church, I offer the following Christmas reflection. Written some twenty years ago and shared with a seemingly countless number of persons, it gives me great pleasure to share the same words with this particular audience.

No other Christian feast has penetrated so deeply into our experience as Christians. Our main experience at that time is that God has said "Yes" to the world. The Lord Jesus did not enter a strange world, but "came to his own home" (John 1:11). This means that our world is not just our own home – and not just our world, but also our whole experience and everything that happens to us; even our very own selves. All this doesn't belong to us alone. God holds sway in everything: a dynamic quality pointing to something beyond our understanding.

What is the basic content of the birth of Jesus, and how should we react to it? On the one hand, it is a message of joy. On the other, it is a call to follow Jesus. We must bear both aspects in mind if we are to think effectively as Christians about the mystery of Christmas. For now this attitude of Jesus' incarnation is something that cannot come to an end. God has given us Christ – a new beginning. Christmas, therefore, is to be more than just a mood. It is, above all, a task that we as Christians have to carry out in our lives. God became one of us and chose to live among us. Ours is a human God, who calls on us simply to be fully human. And so our humanity is to be lived in joy and in imitation. It is not easy to say which is more difficult to achieve nowadays. Yet God calls us to live in joy and to imitate Jesus in our lives. This is the message of Christmas and its summons to us.

This Christmas, may the Lord bless each one of us with His gifts of peace and goodness. And in turn, may we joyfully imitate His love by sharing these blessings with all others... always.

Fr. Dave

Christmas came and went in rather low-key fashion. Were it not for gathering with family members, I'm afraid that Christmas might have been little more than just another day. The very thing that I most anticipate at that time, and about which I am extremely passionate, is the Liturgy. How disappointed I would be Christmas Eve, 2003.

On December 24th, some two thousand people assembled for a 10:00 pm Liturgy. With no high points to speak of whatsoever, I will present the obvious low points of that particular Liturgy:

- Music was minimal and incredibly lacking, to say the least. A few carols were sung as the people assembled. And of course there were the usual hymns (i.e., gathering preparation of the gifts, communion song and concluding hymn). But few, if any, of the parts of the Mass proper were sung.
- The presiding and preaching were every bit as lacking as the music, if not worse. During his greeting and welcome, Father made it clear that he was in a hurry. Apparently, he was to fly out early the next morning for a week long vacation with family and friends.
- No one but the presiding priest communed from the cup. And with the use of many Eucharistic ministers, the communion rite seemed to pass with the blink of an eye.
- Overall, I had the distinct impression that this particular priest is one of those who approaches Liturgy with little more than the old "Get 'em in and get 'em out" attitude.
- One of my family members observed that despite the large crowd, the Liturgy was "done" in exactly 45 minutes.

- Upon exiting the worship space, my niece (23 years of age) turned to me and said: "And they wonder why I don't go to church." To that statement, I responded: "If this is Liturgy, then it's no wonder to me that large numbers of persons, young and old alike, are not going to church. This is a disgrace to Liturgy!"

In contrast to this nightmare experience, it is somewhat encouraging to know that there are glimmers of hope on the horizon, that there does exist places where Catholics are finding themselves nurtured, challenged, stretched and spiritually lifted up. There, Liturgy is accomplishing all that it is supposed to do. Sadly, however, such places of Catholic worship are few and far between. Still, I continue to pray and hope for the dawning of a new day, a day in which life-giving Liturgies are the norm, **not** the exception.

With this in mind, I was prompted to write to the National Catholic Reporter. Although edited for length, my letter did appear in NCR. That letter, unedited, is presented here:

23 January 2004

I wish I could say that I am joyfully anticipating the coming liturgical "reforms." Instead, I find myself simply dreading what can best be described as liturgical **de**forms. Two of the proposed reforms are worthy of mention, here.

First, when the presider prays the invocation "The Lord be with you," the people of God will be asked to respond "And also with your spirit." Hmmm. There's that body/spirit separation still at work among us. To respond with "And also with **you**" (as is the present practice) directs our words to the whole person, not simply one aspect of his or her being. After all, should we not be praying for the entirety of one's being? And if Rome is adamant about this particular change in liturgical protocol, then perhaps the invocation ought to be changed as well, as a matter of consistency. Thus the presider would pray: "The Lord be with your spirit." Oh, if that doesn't sound like nails on a chalkboard!

Second, the proposed reforms would have the congregation pray these words as part of the Confiteor: "through my fault, through my fault, through my most grievous fault" (of course, accompanied by a breast-striking gesture). About this proposed reform many of my family and friends have echoed the following sentiments:

- "I don't think so."
- "Not gonna happen."
- "Can you say **guilt**?"
- "What next? Women must wear veils at Liturgy? Clickers, so that all might genuflect in unison?"

The energy that has been spent on these pending "reforms" is yet another sad commentary on the state of leadership within the church. This was summed up well in a recent issue of NCR, in which one writer said: "They (the bishops) are simply rearranging the deck furniture aboard the Titanic." Using the same analogy, one could also say this: While some are frantically bailing water and others are literally jumping ship, there sit the bishops sipping wine and listening to a lone violinist playing <u>Nearer My God To Thee</u>. And in so doing, I hope it is they (the bishops) who are striking their breasts and praying "through **our** fault, through **our** fault, through **our** most grievous fault."

Soon after welcoming the new calendar year, Bishop Britt rendered his decision concerning my request to return to active priestly ministry. His letter follows…

January 28, 2004

Dear David,

During our meeting on December 10, 2003 I said that I would be in touch with you by late January or mid-February with my decision regarding your request to return to active ministry in the Diocese of Grand Rapids. I had indicated that I needed time to pray, consult and make a decision.

During this past month and a half I have done all three. I spent time reviewing the entire file as well as consulting with people whom I feel could give me helpful insights into the decision I had to make.

In light of these I have decided that I am not open to your request to return to the diocese. I think that you may not realize the depth of feeling many priests have regarding what has happened in the past, and the anger present. If you were to return you would not experience a very warm welcome among the priests with whom you would be living and ministering.

There are responsibilities that I bear in regards to your well being and I accept them and will fulfill those responsibilities. I would ask that you contact Fr. William Duncan to discuss these matters, particularly regarding insurance coverage, if needed.

As I indicated during our meeting, you are free to seek another diocese in which you might be able to serve. If a bishop is interested in having you serve in that diocese, I am willing to grant excardination from Grand Rapids, though I would have to share with him my reason for not accepting you back in our diocese.

May the Lord guide and assist you in the days and months ahead.

With all best wishes, I remain

Sincerely yours in Christ,

Most Reverend Kevin M. Britt
Bishop of Grand Rapids

The bishop's letter arrived on Saturday, January 31st. Late that evening, I called one of my brother priests of the diocese, one who has been both friend and mentor. In the course of our conversation, he confirmed that my request was discussed at a recent meeting of the presbyteral council. He went on to say this: "Dave, the problem is that yours is a prophetic voice. And as you know, the institution doesn't welcome prophets. It never has." Flattered and very much humbled by such a remark, there remained the prevailing sense that

I have been labeled, by some, as either **having** the problem or **being** the problem. Why? Because I publicly criticized the leadership of the Catholic Church in general. I gave my voice to that which is believed and shared by many within the larger church, priest and laity alike. Near the end of our conversation, I said to my brother priest: "Granted, my resigning altogether from priestly ministry was, in retrospect, a most extreme action. At the same time, however, if speaking honestly and truthfully are viewed by the church (or at least its institutional component) as being problematic, then I will be the very first to say that I do indeed have 'the problem.' So much for being that **'priestly, kingly and prophetic people'** which Baptism calls each of us to be."

For the next few days, a great deal of time was given to repeated readings of the bishop's letter and prayerfully reflecting on his words. Out of this, however, I began to speculate on the following questions of concern:

- What recourse do I have in all of this?
- Aside from the presbyteral council meeting, did there occur any 'private' consultation of the priests whose names I had given to the bishop on December 10th?
- Were the laity consulted as part of the decision-making process? I raise this question because I recall specifically asking the bishop, on December 10th, if he would consider soliciting input from the people whom I had served as priest.

Clearly, these questions would need to be addressed to the bishop himself. In the meantime, I would continue to pray, reflect and discern. I would also write a letter, a general mailing of sorts, which has gone out to literally hundreds of persons. That letter is presented here:

5 February 2004

Dear family members and friends,

Greetings. The journey, the struggle continues. Enclosed is a copy of the letter from Bishop Britt, denying my request to return to the diocese of Grand Rapids. You are free to share it with whom ever you so desire.

I cannot say that I am surprised by the bishop's decision, although I am certainly disappointed. I do plan to appeal the decision. However, this very moment is not the proper time for writing such a letter. I will need a few more days to reflect, pray and regroup. Then will I be able to articulate myself in such a way as to convey what needs to be said, as opposed to what I might **like** to say to the bishop.

At this time, as part of my appeal, I am putting out the word and asking you, God's people, to consider orchestrating a letter writing and/or petition campaign on my behalf. Perhaps this is something of a long shot, but what have I to lose at this time? What does any one of us have to lose? Some might ask the question: "Do you think such a campaign will effect a change in the bishop's decision?" To this I would say: "Maybe, and maybe not. **BUT**, it does have the potential for making at least one bishop aware that the issues will remain." I may very well have to go away, when all is said and done, but the issues which led to my resigning in protest are **not** going to disappear, unless of course all of us choose to remain silent.

In recent months, a number of letters have been exchanged between Dr. Eugene C. Kennedy (a Catholic psychologist at Loyola University, Chicago, who shares many of the same concerns as you and I, regarding the church) and myself. In a letter that went out just a couple days ago, I said to Dr. Kennedy: "With all of this being so fresh and very raw, it will be necessary for me to take some time to digest and prayerfully reflect on the bishop's words." I went on to say that as I formulate my letter of response to Bishop Britt, I will do so against the backdrop of a number of concerns. Among them, the following:

- Is there a forum in which I can appeal the decision?
- As with so many dioceses, Grand Rapids has its history of treating alcoholic priests, of tolerating sexual acting-out (between adults, male and/or female) on the part of some of our clergy, of simply reassigning priests who have embezzled parish funds. And of course, there are the years of covering up for and protecting priest pedophiles, with the people of Grand Rapids still having no idea as to the total dollar amount which the diocese has spent on this issue alone.
- In light of the above, I find it truly puzzling that I am left with the feeling that I have been deemed unsalvageable as a priest.
- Where are redemption, compassion and forgiveness to be found in all of this? Are these not at the very heart of all that we preach and teach as followers of the Christ?

In closing out this letter, allow me to do so in exactly the same words that have just gone out to one of my brother priests: In a sense, I feel as though I am literally fighting for my life. The real battle, however, has to do with matters of faith, conviction, conscience and personal integrity. Are these not worth fighting for? And so I ask you, once again, will you consider lending your support and raising your voice on my behalf?

Peace, love and joy in Jesus Christ...always,
Fr. David L. Harpe

As stated, the preceding letter was sent to hundreds of persons. Included in the general mailing were a number of my brother priests who, at the time of my resignation, had called or written to say such things as "we're with you," "you are right on all these issues" or "we support you." To these priests, there was also written a personal note in which I asked two direct questions:

- Did the bishop **'privately'** consult you with regard to my situation?

- When are you going to come forth from the shadows and speak these things in the light of day?

To date, none of these priests have responded affirmatively to the first question. As for the second, not one of the brothers has responded. That I may not be overly harsh toward these good and faithful men, if one but considers the present climate within the church, and therein the very real fear of reprisal, I can understand their reluctance.

In the midst of addressing and stamping envelopes, and out of prayerful reflection, the words came. It was time to write the bishop. Here is my letter of appeal:

13 February 2004

Dear Bishop Britt,

At this time it has been nearly two weeks since I received your letter, denying my request to return to the diocese. As I have stated to family and friends, I am not surprised by your decision. I am, however, disappointed. Naturally so. Having carefully and prayerfully reflected on all of this, I am now writing to see if there is an avenue or process by which I might appeal the decision. Toward that end, I would ask you to consider the following:

- My file, along with the comments of others (written or spoken), provides only a glimpse of who I am. With this in mind, I am left with the impression that the decision-making process centered primarily around my words and actions during a near two-year period of time. It would seem that little consideration was given to the whole.
- Relative to point #1, I am compelled to ask if you consulted, privately, with those priests whose names I had given to you on December 10th.
- Also relative to point #1, I have written to family and friends alike, asking if they would consider writing you on my behalf. Please understand that I have done so not to be

confrontational. Rather, I would like you to see that there is much more to me than simply "the file," either in part or in its entirety.

- To some degree, I can appreciate your assessment that I "would not experience a very warm welcome among the priests," were I to return to the diocese. However, it is my firm belief that those priests who would not give me a warm welcome are the same priests who have never warmed-up to me in nearly ten years. As I said to you on December 10th, "I would return with head held high." And where welcome truly matters, among some of the priests as well as the laity, I am confident that I will find it.

- I do not know which priests are currently serving on the presbyteral council. I am thinking of one in particular, whose name I will not print here. What disturbs me is to know that that person, with his own colorful history of acting-out, might have influenced the decision to deny my request to return to the diocese. If indeed that individual is serving on the council, then it would seem only just that he should have abstained from commenting on matters concerning my wish to return to active ministry.

- Grand Rapids, like so many other dioceses, knows all too well how truly human and even sinful we priests can be. The diocese also has its history of treating priest-addicts, of knowing that there exists sexual acting-out on the part of some of our clergy (between adults, male and/or female). And of course, nothing has served to better underscore our humanity as well as our sinfulness than the sexual abuse of children by priests. In light of all of this, I find it particularly sad and puzzling that I am left with the feeling that I have been deemed unsalvageable as a priest.

It is my hope, Bishop, that our dialogue will continue. In closing out this letter, I am compelled to ask: Where is redemption, compassion and forgiveness to be found in all of this? Are these not at the very heart of all that we preach and teach as followers of the Christ?

Sincerely yours in Christ,
David L. Harpe

Within a short period of time, I would receive the following acknowledgment from Bishop Britt:

March 1, 2004

Dear David,

I am writing to acknowledge receipt of your letter of February 13, 2004. I have read your letter and the points you outline therein.

Though I realize that you are disappointed in my decision, I still stand by it. Should you feel you want to make a canonical appeal to this decision, I would suggest you begin by speaking with Fr. William Duncan or another canonist of your choosing.

May you also have a fruitful Lenten season.

With all best wishes, I remain

Sincerely yours in Christ,

Most Reverend Kevin M. Britt
Bishop of Grand Rapids

At the suggestion of the bishop himself, I have since been in contact with Fr. Bill Duncan, to whom I have expressed my desire to make a canonical appeal to the bishop's decision. Apparently, the process does not move along quickly, which comes as no surprise. And so I wait.

Near the end of his rather short and non-revealing statement of acknowledgment, Bishop Britt said to me: "May you also have a fruitful Lenten season." Although I remain in something of a liminal state of existence, it pleases me to say that Lent has indeed borne much fruit. With its many images and themes (desert, wandering and wondering, letting go, emptying of one self, dying, etc.), Lent assists us well in providing a wonderful opportunity for spiritual reflection and self-examination around this all-important question: What is there within me/us that needs to fall to the earth and die (like the grain of wheat), so that the Lord can raise it up and give it/us new life?

All too often, Lenten preaching seems as if it is pointed or directed only to the people in the pews. In recent years, however, especially in light of the clergy sex abuse scandal, it has become evident that our church leaders should (**must**?) likewise subject themselves to collective spiritual reflection and self-examination that is centered around the very same question: What is there within us (**the institution**) that needs to fall to the earth and die, so that the Lord can raise it up and give it new life?

Throughout Lent I continued to receive calls as well as letters of support and encouragement. These have come from both near and far. In some instances, individuals and/or families have shared with me copies of the letters, which, as a result of my general mailing, they have written to Bishop Britt on my behalf. Some have also shared with me the bishop's written response, which was little more than an acknowledgment of receipt of their letters. To the bishop's credit, however, he has at least given them the courtesy of a response. And for some individuals, it has remained impossible to write such a letter to the bishop, so mixed are their feelings and emotions surrounding the problems that are facing the church. More than a few times have I heard people say, in these or similar words: "I am so angered, disillusioned and outraged with the leadership of our church, I don't even know where to begin writing."

Also during Lent, and with great timing, I read these two newly released books:

- VOWS OF SILENCE: Abuse of Power in the Papacy of John Paul II, by Jason Berry and Gerald Renner.
- Keep the Faith, Change the Church, by James Muller and Charles Kenney.

In light of the present situation (particularly the sex abuse scandal) and given the general climate within our church, I strongly encourage every adult Catholic to read these incredibly powerful, telling and eye-opening works.

Finally, during Holy Week I was most pleased to receive the following note from one of my brother priests of the diocese of Grand Rapids:

Easter – 2004

Dear Dave,

Just wanted to send you Easter greetings. I appreciated your letter in January, but was sad to hear what it had to say.

I wish I knew what to say or how to give you hope. Perhaps this time is your own Calvary. But I do believe it will lead to Easter life, and that is what I pray for you. May the promise of Easter be yours.

Do not lose heart. Be assured of my thoughts and prayers, especially in the mystery of Easter.

Your brother,
Tom

Lent, as we know, is not only a time for journeying into ourselves. It is also the path that leads us to Easter. Much as it was with Christmas, 2003, this Easter was also something of a low-key experience, so difficult it is to find that space or place where Liturgy is truly a life-giving and life-celebrating encounter between the human and the divine. Nonetheless, Easter did come. By way of prayerful reflection and discernment, reading, writing and especially through the tender words of others, Lent proved to be very fruitful indeed. It has led me to an Easter experience that is marked by the wonderful feeling that I have been vindicated with regard to all that I have said, in the company of others, about the church's need for reform and renewal. To many, this might seem small in terms of our usual thoughts about Easter. But at the deeper spiritual level, small or large, this IS an Easter experience. And as such, it must be celebrated. **ALLELUIA!**

Chapter Four
WHERE DO WE GO FROM HERE?

Much has been shared within the preceding pages of this book. There has appeared something of laughter and tears, hunger and thirst, hopes and dreams, joy and sadness, pain and struggle, possibilities and reality. As I contemplated how I might bring this writing to closure, it occurred to me that it is simply impossible to do so. This is not my story alone. Nor is it the story of the people of St. Joe's. It **IS** the story of all who make the spiritual journey. As such, this is a book that is forever in the process of **being** written. From within the Catholic tradition, I have added but a few brief reflections to the larger story of all who travel the spiritual path of Christianity, believing that every Christian – in some way, shape or form – is called upon to be something of a contributing writer. Thus there can be no "period," no closure to the story, at least in this world.

Given what has been written in these pages, especially as pertains to the sadness and unrest that are endured by many within Catholicism, one might ask: "Where do we go from here?" We go to the source, that is, to Jesus Christ and His Word. As witnessed by the lived tradition, in every generation we are continually returning to the source as we journey toward the fullness of the Kingdom. And we do so knowing that if we are looking for Him, we will not be disappointed. He will indeed reveal His presence to us along the way, perhaps especially in those moments when we think Him to be distant or possibly absent. During the Easter season we are reminded of this lesson by way of the following story, which is really our own:

Two of them that same day were making their way to a village named Emmaus seven miles distant from Jerusalem, discussing as they went all that had happened. In the course of their lively exchange, Jesus approached and began to walk along with them. However, they were restrained from recognizing him. He said to them, "What are you discussing as you go your way?" They halted, in distress, and one of them, Cleopas by name, asked him, "Are you the

189

only resident of Jerusalem who does not know the things that went on there these past few days?" He said to them, "What things?" They said: "All those that had to do with Jesus of Nazareth, a prophet powerful in word and deed in the eyes of God and all the people; how our chief priests and leaders delivered him up to be condemned to death, and crucified him. We were hoping that he was the one who would set Israel free. Besides all this, today, the third day since these things happened, some women of our group have just brought us some astonishing news. They were at the tomb before dawn and failed to find his body, but returned with the tale that they had seen a vision of angels who declared that he was alive. Some of our number went to the tomb and found it to be just as the women said; but him they did not see."

Then he said to them, "What little sense you have! How slow you are to believe all that the prophets have announced! Did not the Messiah have to undergo all this so as to enter into his glory?" Beginning, then, with Moses and all the prophets, he interpreted for them every passage of Scripture that referred to him. By now they were near the village to which they were going, and he acted as if he were going farther. But they pressed him: "Stay with us. It is nearly evening – the day is practically over." So he went in to stay with them.

When he had seated himself with them to eat, he took bread, pronounced the blessing, then broke the bread and began to distribute it to them. With that, their eyes were opened and they recognized him; whereupon he vanished from their sight. They said to one another, "Were not our hearts burning inside us as he talked to us on the road and explained the Scriptures to us?" They got up immediately and returned to Jerusalem, where they found the Eleven and the rest of the company assembled. They were greeted with, "The Lord has been raised! It is true! He has appeared to Simon." Then they recounted what had happened on the road, and how they had come to know him in the breaking of bread.

Luke 24:13-35

I believe it goes without saying, but I will say it anyway: We, the church, are in the midst of one of those darker moments of our faith journey. It is also every bit as evident that we must, together, return again to Jesus Christ and His Word. In so doing, we must likewise welcome Him into the depths of our individual and collective hearts. Then will the darkness lift. Then will we find ourselves, once again, as a people on their way. And only then will we see ourselves becoming something more of that new creation which the Lord is ever calling us to be.

Toward that end, to which we are always traveling, and that we might truly **BE** church along the way, I offer these parting suggestions:

- Once more, we must return to the source.
- We must, as St. Paul instructs us, "pray constantly and attentively for all in the holy company." Ephesians: 6:18b And at this particular moment, as we approach the solemn feast of Pentecost, I would encourage all to consider praying not so much for a new outpouring of the Spirit, but rather a new **openness** to the Spirit which has already been poured forth upon us.
- There must be ongoing dialogue within the church. And it is imperative that the laity be very much a part of the conversation. Always.

Finally, as a Christian people we profess Jesus Christ as the Alpha and Omega, the beginning and the end. In this dark moment, as our church cries out for healing, reform and renewal, let us rediscover our sense of direction as well as our very purpose by listening to the One who ultimately has – and IS – the final word:

When they had eaten their meal, Jesus said to Simon Peter, "Simon, son of John, do you love me more than these?"

"Yes, Lord," he said, "you know that I love you." At which Jesus said, "Feed my lambs."

A second time he put his question, "Simon, son of John, do you love me?" "Yes, Lord," Peter said, "you know that I love you." Jesus replied, "Tend my sheep."

A third time Jesus asked him, "Simon, son of John, do you love me?" Peter was hurt because he had asked a third time, "Do you love me?" So he said to him: "Lord, you know everything. You know well that I love you." Jesus said to him, "Feed my sheep." John 21:15-17

APPENDIX

- <u>All Are Welcome</u>, by Marty Haugen
- <u>We Are Called</u>, by David Haas
- <u>No Longer Strangers,</u> by David Haas
- <u>Beyond This Day</u>, by Phil Konczyk

Rev. David L. Harpe

All Are Welcome

Marty Haugen

1. Let us build a house where love can dwell And
2. Let us build a house where proph-ets speak, And
3. Let us build a house-where love is found In
4. Let us build a house where hands will reach be-
5. Let us build a house where all are named, Their

all can safe-ly live. A place where saints and
words are strong and - true, Where all God's chil-dren
wa-ter wine and wheat: A ban-quet hall on
yond the wood and stone To heal and strengh-en,
songs and vis-ions heard And loved and treas-ured,

chil-dren tell how hearts learn to for-give. Built of
dare to seek To dream God's-reign a - new. Here the
ho-ly ground, Where peace and jus-tice meet. Here the
serve and teach, And live the word they've known. Here the
taught and claimed As words with-in the Word. Built of

hopes and dreams and vi-sions, Rock of faith and vault of
cross shall stand as wit-ness And as sym-bol of God's
love of God, through Je-sus, Is re-vealed in time and
out-case and the stran-ger Bear the im-age of God's
tears and cries and laugh-ter Prayers of faith and songs of

grace; Here the love of Christ shall end di-vi-sions:
grace; Here as one we claim the faith of Je-sus;
space; As we share in Christ the feast that frees-us:
face; Let us bring an end to fear and dan-ger:
grace, Let this house pro-claim from floor to raft-er:

All are wel-come, all are wel-come, all are wel-come

in this place.

We Are Called

David Haas

Rev. David L. Harpe

No Longer Stranger

vocal arrangement
by Jeanne Cotter

David Haas

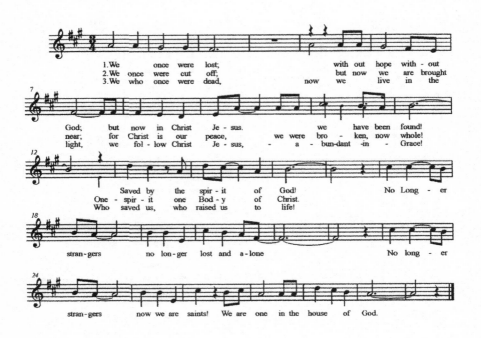

1. We once were lost; with out hope with - out God; but now in Christ Je - sus. we have been found!
2. We once were cut off; but now we are brought near; for Christ is our peace, we were bro - ken, now whole!
3. We who once were dead, now we live in the light, we fol - low Christ Je - sus, - a - bun-dant -in - Grace!

Saved by the spir - it of God! No Long - er
One - spir - it one Bod - y of Christ.
Who saved us, who raised us to life!

stran-gers no lon - ger lost and a - lone No long - er

stran-gers now we are saints! We are one in the house of God.

196

Dedicated to the people of Saint Joseph Parish, Muskegon, Michigan

Beyond This Day

Paraphrased from Psalm 89

Phil Konczyk

Be - yond this day   wher - ev-er you will find me,   I will

sing of the good-ness of the   Lord_____ Wher - ev-er you will find me be - yond this day.

1. Of the promises of God my song shall last beyond this day. Of the faithfulness of God I shall proclaim to all future generations. Established in the heavens is your truth. "Eternal is your kindness" I have heard you say.

2. "With my Chosen one I have made a covenant, beyond this day. This to David my dear servant I have sworn. To all future generations established for all ages is your throne. Certain is my promise" I have heard you say.

3. "And my kindness shall not falter, but remain beyond this day. And my covenant stand firm, it shall not yield. To all future generations who call to me aclaiming, 'You, my God, my rock, my savior and my Father!' I will hear them say."

SUPPLEMENTAL READING

As a means of underscoring that which I have written regarding the spiritual journey and all that is lacking or at least crying out from within Catholicism, I would draw the reader's attention to the following works:

ARTICLES

- Quinn calls again for governance reform, by Dan Morris-Young. This particular piece appears in the following pages, after which this writer offers some further thoughts and reflections.
- Healing The Wound, by Dr. Eugene Cullen Kennedy, of Loyola University, Chicago. This excellent piece can be found by visiting the National Catholic Reporter's online archives: http://natcath.org/NCR_Online/archives2/2003d/100303/100303a.php or by e-mail: ncrad@natcath.org Search for cover story, Healing The Wound, issue date of October 3, 2003

BOOKS

- VOWS OF SILENCE: Abuse of Power in the Papacy of John Paul II, by Jason Berry and Gerald Renner.
- Keep the Faith, Change the Church, by James E. Muller and Charles Kenney.

Quinn calls again for governance reform
By Dan Morris-Young

Curial reform and a hard look at the exercise of papal primacy are critical for the health of the Catholic Church as well as any realistic hopes of future Christian unity, Archbishop John R. Quinn told an audience of 250 here Sept. 8. But it would take "an insistent pope" working with a broadly representative "consultative mechanism" to pull if off, he said.

In an address titled "Shaping a new kind of papal government: a permanent synod," the retired San Francisco archbishop covered many of the themes from his much-publicized 1996 Oxford lecture and his 1999 book, The Reform of the Papacy: The Costly Call to Christian Unity (Crossroad Publishing Co.), including a strong call for decentralization of church governance. "Highly centralized" administrative power in the Roman curia has led to frustration among Catholic church people around the world, he said, as well as keeping other Christian denominations wary about how they would be treated if agreeing to formally recognize Petrine primacy. The former president of the U.S. bishop's conference (1977-1980) said there is global concern about curial conduct, including "no meaningful" local consultation on appointment of bishops, reversing or ignoring actions of episcopal conferences, and meddling in local church issues.

For example, he said, he saw little reason "for Rome to be involved in the remodeling of the cathedral in Milwaukee" or for curia officials to be second-guessing English-language liturgical translations. Speaking with NCR the week prior to his Spokane lecture, Quinn said he is "often pulled aside" by fellow bishops around the world "and told they hope the Vatican takes it (his call for decentralization and reform) seriously."

At the same time, Quinn told NCR, he felt there would be "great resistance" within the Vatican to an examination of how bishops are currently appointed as well as to "placing more responsibilities in the episcopal conferences."

Support, however, for his call for reform of church administration has come from a wide spectrum of church people, he said. "Some very conservative people I know have been quite positive about the book – because they see these kinds of changes are necessary and need to be made."

In his address Quinn advocated a "permanent synod made up of bishops from around the world and with limited terms" that would operate "permanently at the side of the pope" to govern the church. Such a synod, he said, would "be superior to the curia" and reduce the curia's role from a legislative, governing one to an administrative capacity. Interestingly, Quinn was asked by the executive director of Spokane's Catholic Charities, Donna Hanson, how a permanent synod of only bishops might reflect the insights or ideas of church women.

Calling it "an obvious question," Quinn said he advocates much broader inclusion of both lay men and women in church decision-making and governance. In his book he advocates a lay role in selection of popes; he also envisions laity being not only members of Vatican councils, but heads of some as well.

In the structure of the curia there are congregations, councils and tribunals. A creation of Vatican II, councils include organizations such as the Pontifical Council for the Family and the Pontifical Council for Justice and Peace.

"Lay persons have in fact been added to the membership of some councils," he writes, "yet 35 years after" the Second Vatican Council's call for increased lay participation "half the curial councils still have no lay men or women members. Lay persons are not members of any of the congregations."

Quinn told the audience gathered in Gonzaga University Law School's moot courtroom that his book, his Oxford lecture and other addresses have been in large part a response to Pope John Paul II's own invitation to bishops of the world to dialogue with him about the exercise of papal primacy. That invitation is explicitly contained in the 1995 encyclical Ut Unum Sint ("That They May Be One"), Quinn said.

The archbishop said he hand-delivered a copy of his book Reform of the Papacy to the pope on Dec. 6, 1999, and then went "down the street" and gave one to Cardinal Edward Cassidy, then head of the Pontifical Council for Promoting Christian Unity. Quinn said Cassidy made it clear Ut Unum Sint was the personal work of the pope himself, not a document primarily prepared by a committee or commission and then signed by the pontiff.

Cassidy told Quinn the pope had given a copy of Ut Unum Sint to him for his pontifical council to review. "He thought the pope wanted him to curialize and vaticanize it," quipped Quinn.

Almost immediately after Cassidy delivered the reworked Ut Unum Sint text to the pope, he received a phone call from the pontiff, Quinn said. "What you have sent me is not what I wrote," the pope reportedly told the cardinal.

John Paul II's personal investment in the encyclical makes its call for conversation about exercise of the papal office all the more significant, Quinn said. He called Ut Unum Sint "revolutionary" and "the Magna Carta of ecumenical efforts in the future."

Quinn said the pope makes it clear in the encyclical that he sees the first millennium as a model for possible changes in how the papacy is exercised. Ut Unum Sint "locates the primacy of the pope within the episcopate, not something outside and above it," Quinn said, and does not portray it as "a sovereign, monarchical office."

In the encyclical the pope alludes to fellow bishops as "brothers," Quinn pointed out, "not sons."

The pope "focuses strongly on the first millennium papacy," Quinn said, emphasizing that the "sketch" or "paradigm" of papal authority in that time of church history rested on a "synodal principle" – a communion of churches that existed on their own, at the same time being united "in communion" with the bishop of Rome.

Characteristically, Quinn said, the pope in that era would not intervene or take part in local church questions, only exercising a role at times of "major issues."

While John Paul II appears to be open to a papacy whose authority is wielded in a more consultative, collaborative way, he is also clear he would not accept the idea of a papal office that verges on titular, Quinn said. John Paul II insists the papacy must have "true authority" to properly fulfill its function, Quinn explained.

By indicating it is "not necessary for the papacy to be exercised through a strong, centralized authority" and by acknowledging the "synodal principle," the pope has opened the door to serious discussion of Christian unity, Quinn said, notably with Orthodox churches. The Orthodox and others, said Quinn, are amenable to the concept of papal primacy. It is how that office currently functions that concerns them.

For example, he said, in 1999 the Anglican-Roman Catholic Dialogue issued a statement on the possibility of the primacy being recognized even if the two churches were not in full communion.

In a statement earlier this year, he added, the World Council of Churches said it was "open to continued study" on the topic of papal primacy. Quinn repeatedly called the Vatican to allow great autonomy for and collaboration with episcopal conferences. It often appears today, he said, that bishops are simply "field managers" carrying out dictates of the curia.

In Reform of the Papacy, Quinn illustrated his concern about curial "micro-managing" with an anecdote about his reception of an honorary doctorate in theology from a pontifical school. To be allowed, the degree had to be approved by the Vatican Secretariat of State, the Congregation for the Doctrine of the Faith and the Congregation for Catholic Education. "It would suggest," Quinn writes, "that a bishop in communion with the pope could be rejected as a candidate for an honorary degree yet would have the qualities necessary to be a teacher and witness of faith as a bishop."

Although he has received little direct reaction to his book from the Vatican, Quinn said he knows officials there are aware of his message, including Cardinal Joseph Ratzinger, prefect of the Congregation for the Doctrine of the Faith. In a book-length interview with the cardinal, journalist Peter Seewald quotes the high-ranking curia official as alluding to Quinn's book when questioned about the burdens and vast complexities of the papal office.

"Retired Archbishop Quinn of San Francisco has vigorously argued for the need for decentralization. Certainly, much can be done in this area," Ratzinger is quoted as saying (NCR, Sept. 28, 2001).

Quinn's address was sponsored by the Spokane diocese, Gonzaga University, and Catholic Charities' Parish Social Ministry Office.

Personal Thoughts / Reflections

I am most appreciative of Dan Morris-Young's article, <u>Quinn calls again for governance reform</u>, as well as the writer's permission to include the piece here. At the same time, I applaud the courage and conviction of Archbishop John Quinn. However, I am compelled to share a few thoughts and/or observations concerning the matter of governance reform within the church.

First, there is that which is painfully obvious. It has been nearly nine years since the release of the papal encyclical, Ut Unum Sint. Now as then, not much (if anything) has changed. It certainly appears that there is a glaring contradiction between the words and the actions of the Holy Father.

Secondly, I would like to pose a challenge to the retired archbishop, which, indirectly, is a challenge to those who are currently serving as bishops in the church: The next time a fellow bishop says "they hope the Vatican takes it (Quinn's call for decentralization and reform) seriously," perhaps the archbishop would consider responding in this or similar fashion: "**YOU** are a bishop, a 'brother,' one among equals. **YOU** need to develop the intestinal fortitude or, for lack of a better expression, the b---s to be a bishop, daring to faithfully speak out in truth."

There is the challenge for the few. For the majority of bishops, however, they will remain content to stay the course of this "business-as-usual" approach to church, yet occasionally complaining and grumbling among themselves. But when it really matters, when they have an opportunity to speak out, to call for reform and renewal within our household of the church, they will be silent.

Finally, something of an irony: On September 29th, just three days after my having read Morris-Young's article, the news media announced that the pope had created 31 new cardinals. Are we to presume that there was any "meaningful local consultation" surrounding these 31 appointments/elevations? I think not. Aside from possibly consulting the good-old-boys network within the Roman curia (for the purpose of tapping into the pool of office boys and ladder-climbers who are most likely to pledge their blind obedience to the pope, in the name of "orthodoxy'), the Holy Father alone continues to choose **who** is to become a bishop or cardinal, as well as **where** that individual will exercise his episcopal ministry. Tell me, where is the "meaningful local consultation" in this process?

ACKNOWLEDGMENTS

As I drew close to completing this writing exercise, I came to the rather profound awareness that this is as near as I will ever come to experiencing pregnancy and the birthing of life. Deep within, something of a new creation was conceived. This was followed by months of joyful anticipation, though not without occasional moments of pain, discomfort and irritability. All the while, however, there has been a prevailing sense of excitement, of longing to see that which has been growing within me. Then, the appointed time arrived. With a few deep breaths and a couple of last-minute pushes, birth came easily. Now I am able to see it, touch it and literally cradle this new life in my arms.

This book, this "birth," is not of my own volition. No, it has come forth from the hearts, love and faith that are shared by many. I, for whatever reason, am simply the one who has enjoyed the privilege of giving written expression to **our** story. Even this might have proven impossible, were it not for the presence of others along the way. As such, I must extend the following words of gratitude:

- To our creator God, in whom all things are truly possible.
- To my parents, Ben and Joann, for handing-on to me the gift of their faith.
- To the women, men, young persons and children who continue to inspire me by way of their faith and their desire to live that faith in joyful imitation of Jesus Christ.
- To the late Joseph Cardinal Bernardin: Though he was somewhat small in stature, he was and will always remain a towering figure within the church.
- To the late Fr. Gorman Sullivan, who taught Sacramental Theology at Mundelein Seminary: The greatest gifts he imparted to his students were 1) the need to "get this stuff into our blood, that we may live it," and 2) the constant challenge to dream and theologize BIG about ourselves, our faith and, ultimately, our God.
- To family and friends, Catholic and non-Catholic alike, who continue to pray for me, support me and at times even carry me with their undying love.

- To all who find themselves pushed to the edge, ignored, or treated unjustly within our church.
- To my brother, Gerald: It was at his home where I first began writing this story well over a year ago.
- To another brother, Ben, and his wife, Georgiann: They opened the door to their home, which, in recent months, afforded me the space needed to bring this work to completion. Without them, I can only wonder if I would now be writing these words.
- To Vince and Carol Pharries, as well as their son, Ben: Many of these pages were first copied at their place of business, at no cost to myself.
- To Dan and Deb LeMire, along with their daughter, Jessica: Not only have they, like so many others, become my extended family, but they have also provided great care and a lot of food for my cat, Jeter, during my time away.
- To Shirley Richardson, whose skill and generosity have transformed an old photo of St. Joseph Catholic Church, Muskegon, into the cover page of this manuscript.

Thanks to all of you. And thanks to all persons who travel this path together, who cry out for spiritual nourishment, who continue to call for reform and renewal within our church.

Scripture passages were taken from THE NEW AMERICAN BIBLE, 1970, P.J. Kennedy & Sons – New York, New York

ABOUT THE AUTHOR

As a priest, the author has an interest in and a passion for the spiritual journey not only of himself, but especially the very people with whom he makes that journey as a servant-leader. Outraged by the present leadership of the catholic church, which appears near-totally divorced from the people whom that body is likewise to lead and serve, this priest publicly resigned from active ministry as a statement of protest.

Shortly after resigning in January 2003, the priest returned to a writing exercise which was begun two years earlier. The result is this book, <u>Holy People, Holy Irreverence: A Church In Need of Reform and Renewal,</u> in which the author reflects on the personal and communal journey of faith, a growing spiritual hunger within the church and the failure of an institution to satisfy such hungers.

Made in the USA
Monee, IL
26 April 2022

95483521R00132